Henry Demarest Lloyd

Labor Copartnership

Notes of a Visit to Co-operative Workshops, Factories and Farms in Great Britain

and Ireland, in which Employer, Employé, and Consumer Share in Ownership,

Management and Results

Henry Demarest Lloyd

Labor Copartnership
Notes of a Visit to Co-operative Workshops, Factories and Farms in Great Britain and Ireland, in which Employer, Employé, and Consumer Share in Ownership, Management and Results

ISBN/EAN: 9783337322861

Printed in Europe, USA, Canada, Australia, Japan

Cover: Foto ©berggeist007 / pixelio.de

More available books at **www.hansebooks.com**

LABOR COPARTNERSHIP

Notes of a Visit to Co-operative Workshops, Factories
and Farms in Great Britain and Ireland, in
which Employer, Employé, and Consumer
Share in Ownership, Management
and Results

BY

HENRY DEMAREST LLOYD
AUTHOR OF "WEALTH AGAINST COMMONWEALTH"

ILLUSTRATED

"*Labor is the house that love dwells in*"
RUSSIAN PROVERB

NEW YORK AND LONDON
HARPER & BROTHERS PUBLISHERS
1898

BY THE SAME AUTHOR.

WEALTH AGAINST COMMONWEALTH. 8vo, Cloth, $2 50. Popular Edition, Half Cloth, $1 00.

It is not a book of an hour, nor of a day, but of a life, and no one who examines it for the first time, or for the hundredth, will leave it without a new inspiration and a broader view of the future of humanity, and of his responsibility to it.—*Brooklyn Standard-Union.*

It is a remarkable, an amazing story. Altogether we may say that a stronger indictment of the present dominant forces in industry has not been put in print.—*Springfield Republican.*

NEW YORK AND LONDON:
HARPER & BROTHERS, PUBLISHERS.

Copyright, 1898, by HENRY DEMAREST LLOYD.

All rights reserved.

CONTENTS

I. THE NEW WORD AT THE DELFT CONGRESS	1
II. ON THE LAND	9
III. SOME COPARTNERSHIP FARMS	29
IV. GOLDEN VEINS OF THE EMERALD ISLE	52
V. GROUND RENTS OF BROTHERHOOD	80
VI. FROM CAPITALISM TO CO-OPERATION IN LEICESTER	100
VII. KETTERING	121
VIII. LOSS-SHARING	154
IX. MUSHROOMS THIRTY YEARS OLD	166
X. WORKMEN DIRECTORS IN A GREAT COMPANY	191
XI. THE WORK OF GENERATIONS	214
XII. THE NEW JOY OF LABOR	235
XIII. WORKMAN VERSUS BUSINESS MAN	251
XIV. TWO KINDS OF STORE DEMOCRACY	273
XV. THE SCOTCH THISTLE	299
XVI. SET ON A HILL	318
XVII. BEGINNING AT HOME	326

APPENDIX

1. LABOR COPARTNERSHIP SOCIETIES	337
2. A LABOR COPARTNERSHIP INSTANCE	340
3. CO-OPERATIVE SOCIETIES IN THE UNITED KINGDOM	340
4. THIRTY YEARS OF THE SCOTTISH CO-OPERATIVE WHOLESALE	341

CONTENTS

5. THIRTY-FOUR YEARS OF THE ENGLISH CO-OPERATIVE WHOLESALE - - - - - 341
6. STATISTICS OF THE ENGLISH CO-OPERATIVE WHOLESALE - - - - - - 342
7. TRADES-UNIONISTS AND LABOR COPARTNERSHIP 342
8. ELIGIBILITY OF EMPLOYÉS AS OFFICERS - - 342
9. CHARGES OF SWEATING; THE WALSALL PADLOCK MAKERS - - - - - - 343
10. SOME OF THE STATISTICS OF 1897 - - - 343
11. INDEX - - - - - - - 345

ILLUSTRATIONS

THE SEASIDE HOME NEAR GLASGOW	*Frontispiece*	
WHERE THE "EQUITY" SHOE WORKS STARTED IN 1887	*Facing p.*	108
"EQUITY" SHOE WORKS IN 1894	"	108
LEICESTER'S WORKINGMAN'S HOUSE	"	112
"EQUITY" ROAD AT LEICESTER	"	114
THE KETTERING SHOE WORKS	"	132
KETTERING CLOTHING FACTORY IN 1893	"	142
OPENING OF THE NEW KETTERING CLOTHING FACTORY IN 1895	"	142
CO-OPERATIVE HALL, KETTERING	"	148
THE PICTURESQUE BEAUTY OF NUTCLOUGH GLEN	"	170
HEBDEN BRIDGE FUSTIAN WORKS	"	174
LONDON CO-OPERATIVE BUILDERS AT WORK ON THE CO-OPERATIVE BAKERY AT WOODGREEN	"	238
THE AVENUE, TOTTENHAM, LONDON	"	340

CHAPTER I

THE NEW WORD AT THE DELFT CONGRESS

Last fall I went through Great Britain to see what was being done in the field of production by the co-operative societies. I saw again the now familiar marvels of the distributive stores, which in one generation have grown to a membership representing one-seventh of the British population — and that the picked seventh — and doing a business, manufacturing included, of $272,000,000 a year, with a bank of their own with deposits of $16,000,000, and turning over $200,000,000 a year; and I saw the huge factories which the English Co-operative Wholesale Society has established in its function of capitalist, and which it is avowedly operating on capitalistic lines as to its employés, giving them none of the profits and none of the control.

But there was something else to be seen. There are factories, workshops—and there are now coming to be even farms—planned, set up, operated, and managed by workingmen's brains, money and morals; and that not capitalistically, but co-operatively.

The co-operation of these workingmen is not for themselves alone. Capital takes its place as a wage-earner along with labor, and both, after receiving their earnings—interest for the one, wages for the other—share in the profits or losses; both share, also, in the ownership and management. Even the consumer is recognized as one of the constituents, and shares in the profits he brings, and can share in the control by becoming a stockholder.

Many of the establishments are handsome enough, with all the modern machinery of their trade and the most approved appliances for the health and comfort of the "members," which is the new word for "hands." They have been in successful operation for many years—some of them for as many as twenty and thirty. They are increasing. They now number more than one hundred and fifty, have an aggregate capital of over $5,000,000—as much as the great wholesale store at Manchester has accumulated in its thirty years of growth—annual sales of $10,000,000 and rising, and annual profits of $500,000.

Here the workingman is found playing the President of industry, and playing it well. The aspiration of leaders of the working classes that they should become the owners of the means of production is being realized, step by step, along the lines of least resistance. Industry is democratizing itself, pending the political regeneration of the whole world at once, and labor is capitalizing itself. Archbishops have given up the application of the Golden Rule to business as impracticable; but here it is being attempted by workingmen.

"We must make men as well as money," and "We must help our brothers," is their daily bread. Every dollar of profits, before it is divided, must first pay its contribution to the fund for schools, literature, lectures, to help spread this gospel of self-help by each-other-help, which these simple people do not think themselves smart enough to get taught out of endowments made by pirates. There is no woman question. Women can be seen at the workbenches who sit as Directors in the Board of the factory in which they are wage-earners. Some villages are already almost wholly co-operative; able, practically, to vote unanimously anything they desire whenever the day comes, in their judgment, for a co-operative politics; and the co-operative census is swelling rapidly everywhere.

This was news to me; and as I found that even among distinguished students of social science, leaders of thought, prominent trades-unionists and agitators, and men of affairs in England, there were not a few who were unaware of what was going on under their own windows, it occurred to me that it might be news to some of the people in America. I have taken pains to note the shortcomings, failures, criticisms of the movement. I have referred readers to fuller information of these where space forbade my giving it, and have not concealed the fact that the subject is one of controversy and opposition, even within the co-operative world itself.

This opposition is a latter-day outgrowth of the wealth that has come to co-operation on its commercial side; it is not part of the original creed or deed. In the beginning, in the days of the Christian Socialists, and, before them, of the Rochdale pioneers, it was the co-operative ideal, and has been given the first place in their faith by all the great names—Owen, Kingsley, Maurice, Vansittart Neale, Thomas Hughes, Ripon, Ludlow, Holyoake, Godin, Leclaire—that the workingman should attain self-employment and self-government, sharing in profits and management as a right, not as a favor; and that he should be capitalist, and that there should be a revival of business which shall be the revival of religion. The associations described in these pages are the newest development of this oldest doctrine of co-operation.

The new word at the International Co-operative Congress held in Delft, Holland, in 1897, was, I found, "Labor Copartnership." It was represented by a group of zealous men—"great advertisers," as their opponents smilingly expressed it—who were about the only active propagandists in evidence in the Congress. Except for the desire of a number of the French delegates to speak all at once, and for a few amiable differences of opinion as to minor points of policy, like that of endorsing the proposed International Co-operative Store at Paris,—which has since been opened in a spacious room at 37 Rue Vanves, Paris—the only novelty was furnished by the group among the English delegates who, in the Congress and out of it, were preaching their gospel of labor copartnership.

They piqued interest by refusing to allow their new doctrine to be given any formal definition in the report of the Congress. Their opponents chuckled at their expense at this spectacle of a sect refusing to go on record with a formal statement of just what its doctrine was. But the copartnership people let them laugh; and let it be understood that they were perfectly willing to bear this banter for the moment rather than run the risk of having their movement wrapped too soon in the cramping ligatures of a creed, and measured by doctrinaires for its coffin. The other matters discussed by the Congress were mostly the familiar questions of profit-sharing, agricultural associations, Schultze-Delitsch banks, but labor copartnership was new, and its freshness, as well as the devotion of its advocates, made it one of the catchwords of the occasion, to be quoted in speeches and to be inquired about in private talk.

There was fight, hope, purpose—the spirit of the pioneers, a breath from the people itself—in the delegates who stood for labor copartnership; and when they spoke, they always won applause from the Congress. There was a distinctly different note in their philosophy from that of Godin, Leclaire, and the living successor of these, the distinguished philanthropist of Holland, Van Marken, the yeast and spirits manufacturer of Delft, under whose auspices the Congress assembled. The things these men had done, and admirably, were done for workmen; but labor copartnership brought us account of things done of, by and for workmen themselves. It was a record of self-initiative, and so of a higher promise than these of philanthropic initiative, beneficent as they were.

There is a copartnership community not far from where the Congress sat—at Nunspeet, a settlement where workingmen are gathered together by a man of wealth, M. Molijn, in what it is hoped to make a co-operative commonwealth in miniature; and at Delft, itself, there is a co-operative printing company and a workingmen's village. These have been so organized by Mr. J. C. Van Marken that the workingmen in the printery receive a share in the profits of the employer, and the tenants in the houses a share in the profits of the landlord. This is not in money, but in the shape of an accumulating proprietary interest in the printing-presses, type, etc., of the one, and in the houses of the other. This is to make them, ultimately, owners of their means of production and of their homes in place of the original proprietor, whom they will have expropriated by a scheme originated by himself.

These were things that were being done for the workingmen by employers of good will; but from England the Congress heard a story of productive enterprises, including farms, started by workmen for workmen, owned by workmen, and managed by workingmen and workingwomen. I had come over to attend the Congress, intending after it to go to Germany to look at the People's Banks, in which I had become interested by the fascinating book about them by Henry W. Wolff,* and observe the methods of the Socialist Labor Party; and then to go to Switzerland and see in practical operation its Initiative and Referendum, and the other achievements in self-government which fairly entitled the Swiss to be considered to-day the most successfully democratic nation in the world. But this business of labor copartnership was news; and news is what we will always take first. I saw that, although I had read a good deal about the English co-operative movement, I had failed to grasp its most significant development. I had been warming myself by the picture of a fire. It would be better to go to the fire itself. I, therefore, changed my plans and went to England, to see for myself what this new word of co-operation meant.

* People's Banks, published by P. S. King & Co., London.

CHAPTER II

ON THE LAND

I began my journey in the field of co-operation at its newest, least successful, hardest, and most encouraging point—farming. In connection with an exhibition of their products by the co-operative societies of the Midlands, at Rugby—Tom Brown's Rugby—September 18th, there was a conference of delegates under the auspices of the Co-operative Union to consider specially the possible developments of co-operative production in agriculture. The fine assembly-room of one of the Rugby schools was given to the delegates for their meeting by the Rev. Dr. James, head master of Rugby. An earnest audience of men and women from the societies of Leicester, Kettering, Rugby, and other places in the Midlands, was in attendance. Mr. Henry Vivian, the leading lecturer of the Labor Association, which has been organized to help educate the co-operative constituencies of Great Britain to go into production, but not capitalistically, gave an address;

and then there was an "experience meeting." Every important society in the Midlands that had tried farming, or that wanted to, was heard from. The failures and difficulties were confessed with frankness; and, with equal frankness, the determination to overcome them and succeed.

The feeling about the evils of the land situation was deep—very deep; I do not think it would be exaggerated if called passionate. The room rumbled with approval as speaker after speaker spoke of the "scandal" and "disgrace," and almost uncountable economic waste. But when a man who had been one of the warhorses of the land agitation of the last generation—he had been one of those who rescued for Nottingham the great tract of commons which is now its most precious municipal possession—delivered a fierce tirade against the landlords, he was listened to with evident impatience. All that he said was true, and it was well said; but his audience had got far beyond that stage. What they wanted most to hear of was: "What shall we do about it?"

As I found all through England to be the fashion of co-operative meetings, whether business or popular, the subject in hand was discussed with a constant reference to the interests of others as well as those of self. At this

Rugby meeting, the misery of the agricultural laborer and the help that should be given him was always on the carpet. His low wages and the degradation of his lot were described by nearly all the speakers, and always to sympathetic listeners. "There is something radically wrong about a land system," said one of them, "in which the landlords live in luxury, the farmers (tenants) in comfort, and the laborers in misery." And the hall echoed with "Hear!" "Hear!" In the Midlands, a garden spot if ever there was one, the farm laborers, it was stated, get only 10, 11 or 12 shillings a week, and lose their time in bad weather.

"He has no savings to put into the land," said one of the delegates. "We who have the capital must help him to a better condition, bringing the land to him, and so help ourselves to a higher commercial prosperity." All agreed that one of the first things to be done when the co-operators began to farm was to raise the laborers' wages. This I found a feature of the political economy of co-operation, even among those who did not accept the ideas of labor copartnership—that co-operaters must figure not how little but how much they can pay labor. Even the officials of the Manchester English Co-operative Wholesale Society, and others who oppose the plan of giving the employés a share in profits and management, will always be found to point with pride to the fact that they pay better wages than outside employers, and give benefits not accorded elsewhere. Co-operators believe that poorly-paid men do poor work. "The agricultural laborer cannot do his best work," was said at Rugby, "because he is not paid to do it. He cannot put his heart into his task."

Lack of capital is one of the chronic complaints of the farmer and the laborer the world over. It was a sensation to sit in a meeting composed almost wholly of workingmen considering plans for their "emancipation," and hear the serene confidence with which they declared themselves to have all the capital they wanted. The workingmen in co-operation have become great capitalists. They do not feel that there is any enterprise at which they need to stop for the want of money. They are ready now to meet landlord or capitalist on his own ground. "We have a tremendous amount of money," said one. "We need," said another, "all the commodities that can be produced on the land; we have a mouth ready for every mouthful grown; we have surplus capital; we need only educate the agricultural laborers and pay them living wages, and we shall have our pick of laborers." These societies, like that at

Kettering soon to be referred to, which practice the principle of copartnership, usually begin their preparations for taking up farming by doing propaganda work among the farm laborers of the vicinity.

It was evident here, and at every other gathering of co-operators I attended, that the sentiment was overwhelmingly in favor of radical political action on the land question. Some of those who demand political remedies look upon co-operation as a rival, if not an opponent. But this jealousy, from what I saw and heard among the co-operators, is misconceived. It would be safe to wager that a larger vote will be got for the most radical remedies from co-operators than any other class. "If any man will do his will, he shall know of the doctrine." But as co-operators, these people confine themselves to co-operative methods. There was much complaint of the attitude of the owners of land towards co-operators. Many charge them extortionate rents. In one case mentioned, a society was charged a rent of $60 an acre, and that for land which was two miles out of town. This was at Nottingham. Another instance was given where (near Peterborough) $250 an acre was charged for grass land bought. "They are always talking," it was said, "about how depressed agriculture is;

but they always know how to charge enough for the land." Some of the landlords refuse to let co-operative societies have land on any terms. Others will let it only with the most burdensome conditions, taking from the tenant any discretion to alter the rotation of crops or the methods of culture, even though revolutionary changes are occurring at present in the agriculture of the world. Societies are held down by lease to the old-fashioned succession of crops practiced before foreign grains began to flood the British markets; and have to raise wheat and oats year after year at a loss, when profit could be made in dairying and market-gardening.

Naturally, this tyranny and extortion have created an accumulating public opinion among the co-operators, as well as the people generally, for summary and immediate abolition of this father of all the monopolies, and this feeling found vigorous expression in the conference, as it always does when congress or conference of co-operators discuss the land question. But it does not draw them away from the pursuit of their own special co-operative way of dealing with the problem. Co-operators are resolved not to be switched off from what they can do in their own field by any suggestions of what might be done in some other field. Those who showed that no land at all could be got in their locality, or that the expenses of transfer, the heavy rents and tithes, the railway overcharges, and other things, made general success impossible unless there were sweeping reforms by the interference of the state, got a hearing of the most sympathetic kind, but were always told, in effect, to stand aside.

Co-operators will certainly give an overwhelming support to such reforms when brought forward, but the genius of their own movement is to do what can be done by themselves in voluntary associations, and they are faithful to it. In this, many of them believe they are playing the game of the very political reformers who profess so much contempt for their efforts. History teaches, they point out, that every great change in the political structure had to be preceded by preparatory change in the domestic structure. This latter is the conscious task of co-operation in the social transformation now going on. How much self-government in England and America owes to the voluntary democracy of the Separatists, Congregationalists, Puritans, Wesleyans, who drilled the people in the duties of making their own rules and electing their own rulers, political philosophers like Borgeaud have often told us. In learning to associate in congregations, people prepared themselves to become members of constituencies. In discussing co-operative farming before the Co-operative Congress of 1895, Mr. McInnes, of Lincoln, said most interestingly on this point: "The system of church government devised by the administrative genius of Wesley was, although not designedly so, the first avenue through which the most thoughtful of the agricultural class became familiarized with the principle of acting together, and were brought to realize that they were small, but necessary parts of one great living organization. The silent part which the constitution and procedure of the various Methodist bodies has enabled them to play, in paving the way for the peaceful social revolution which is taking place in the villages, is rarely recognized."

Co-operators are firmly resolved to prove that their method can deal successfully with the land. While Land Restoration and Land Nationalization societies are working to restore and nationalize all the land of Merrie England by a grand stroke of popular sovereignty, the co-operators proceed to restore and colonize, and privately democratize all they can on their own account. They are so completely in earnest in their land campaign as to begin to march at once. They will not wait till land taxes have been reimposed on the landlords, or the Australian system of easy and cheap transfers has been adopted; but will go right on now, taking things as they are, and will get such land as they can, and will prove that it can be cultivated profitably by the co-operative method. As one of them said: "This land question is the most difficult of all. It is our duty as co-operators to put our hands, and heads, and hearts together and lift it up." The same sentiment came from the other side of the social world in Lord Roseberry's speech at the Glasgow Congress. "Until co-operation," he said, "has successfully dealt with the land, its work and its mission are incomplete." Every step co-operators take will help the wider reform; for it will prove that people can farm as people, and it will be getting ready for democratic service on the people's land, when that comes, a body of men skilled both in managing the land and the people.

Co-operative audiences listen with unstaggered cheerfulness to the stories of failures in their farming, and not a few of them were told at this conference. But "failure" has no horror for these people. They have tried failure and got to success—brilliant and glorious success— by the help of failure. You could not persuade an English or Scotch co-operator that anything which anybody could do could not be done still better if the "co-ops" took hold of it. There is no such word as "failure" left in the vocabulary of a movement which, beginning with the tuppence a week of underpaid workingmen in garrets and ridiculous little shops in back streets, has in thirty-six years done a business of $4,500,000,000, and divided among the working people $360,000,000 of money in dividends, opening windows of hope into thousands of lives out of which hope had been taxed by the greed and cruelty of power.

The returns of co-operative farming published in the report given to the last general Congress, that at Perth, in 1897, though accompanied by the comment that they were "certainly far from encouraging," are really not depressing when it is remembered that the movement is still in its first stages, and that the co-operative farmers are "buying experience." In the operations of forty societies for 1896, there were profits of $16,510 and losses of $19,255. The acreage was 4,869, and the capital $499,595. This loss of $2,745 in all as the net result of the experiments of forty societies, with a capital of half a million, in a new business, brought no despondency to any co-operator that I saw or heard of. For the five years, 1892–1896, the total results of co-operative farming showed losses of $65,-265 and profits of $61,465, a net deficiency of only $3,800—certainly a trifling initiation fee; and so it is looked upon by the sentiment of the movement.

This exhibit, it must be borne in mind, is made under the severest system of bookkeeping known to any business. The co-operators carry no bad debts on their books as assets, and force no balances; and they write every year a large depreciation off the value of their property. Every acre in these accounts has been charged with interest on its cost, sometimes with depreciation. Even if the land is really building land, cultivated only *ad interim*, it is charged full interest on its value as such. The farm of the Woolwich society cost about $600 an acre, as it lies just outside of the town and in the path of its growth. It is only a question of a short time when the society will cut it up into building lots for its members. But, meanwhile, the farm has to pay interest on what it cost as building land, not on its value as a farm.

An examination of statistics of co-operative agriculture in detail discloses that the total of losses is due to the very heavy deficit made by the Scottish Wholesale Society's farm, $11,400 in 1896, which was turned into a profit in 1897. This cannot be charged to difficulties inseparable from either Scotch or co-operative farming, for other Scotch co-operative farms made gains in 1896, as shown by their official returns. Dunfermline made $3,125 on 504 acres; the Scottish Co-operative Farming Association, $1,315 on 748 acres. Leaving out the Wholesale, twenty-one British societies have a profit; only thirteen a loss. Their net profits total $16,510; the losses, $7,855. These are the figures given by the Central Co-operative Board in its annual report to the Congress of 1897.

If the increased value of the land were taken into consideration, losses in most cases would also turn into profits. The land co-operators get is often very poor. The land of England is running down. That of the Lincoln society was described as "a rubbish heap." What figures as a deficit in these accounts is in reality frequently the profitable investment that has reclaimed and often more than doubled the value of the land. But co-operative bookkeeping never marks up the value of investments. "We have never appreciated," says the History of the Origin and Progress of the Royal Arsenal Co-operative Society, "the nominal value of any land above its original cost, even when, as in the case of the farm at Bostal, the value of such land since its acquisition by the society has risen immensely."

Stories of success and failure were told at the Rugby conference with equal frankness. Some of the farms referred to were merely co-operative; some were also copartnership. The farm at Lincoln was started in 1889 by the purchase of twelve acres. It converted its "rubbish heap" into a fruitful garden. The profits in 1893 were $75 an acre. In 1897, it reported that for 1896 it had made $780 profit on its fifty-four and one-half acres. Mr. D. McInnes, a delegate from the society, gave this account of its history in the Congress of 1895:

"It grows produce for an assured market, namely, its town membership, but the twelve acres of land which it purchased six years ago were only of poor quality. Good tillage and manuring has altered this, and meanwhile, averaging the whole period, the venture has paid well. But it has done so simply because there has been a mouth ready for every mouthful grown, and for all that human mouths could not take the mouths of a herd of pigs have always been available. Thus all waste has been avoided, and what were formerly partly waste products from mill, store, and abattoir, have been utilized. As co-operation is in its very essence a system of avoidance of waste, co-operators should, wherever possible, organize to use their magnificent and growing resources to develop those of the soil. If our history for the past twenty-five years has not yet given us faith enough to proceed further, we are slow, indeed, to learn—sceptical beyond measure, and there is nothing for it but to wait, to educate, and to organize; for the extension and development of our movement along sound lines depends not on the intelligence of the few, or the faith of the few, but on the faith and the intelligence of the many."

The outcome of this successful experiment by the Lincoln society are these facts, as given by Mr. William Campbell, at the Sunderland Congress: They have realized a good profit, besides charging interest on capital; they have supplied themselves with some of the prime necessaries of life of the finest quality and in the best condition at ordinary prices. They have so improved the quality of the land that it is worth many pounds an acre more to-day than when they bought it. They have paid more for labor on twelve acres of land than ninety-nine farmers out of a hundred pay on fifty, and, by becoming their own producers, have completely annihilated the middleman.

The Newcastle society has done well. A member in the Congress of 1895 said its farm had two hundred and fifty acres of land—one hundred and eighty arable, seventy in grass. They had lately set up a dairy. The store was their market for produce, and the milk was appreciated, for they did not water it. They employed five men, none receiving less than twenty-five shillings per week, and working fifty-six hours. They had worked the farm three years and cleared nine and one-half per cent. That year it was hoped there would be ten per cent profit. But for the five years up to 1897, on $58,640 capital, the profit was only a little over one per cent. This comparatively poor result was largely traceable to the prevalence of an epidemic disease among the cows, to the position of the farm, and to the adverse agricultural conditions of the past few years.

I am indebted to Mr. H. Reed, the secretary of the Ipswich Co-operative Society, for some particulars, under date of December 26, 1897, of the results of their farming. They cultivate sixty acres, purchased ten years ago for $15,000. The society lost on wheat, and now grows vegetables and root crops for its cattle. They have twenty cows and one hundred pigs, and find them a source of profit. The land is poor and wet; but for the year ending October 12, 1897, after charging interest at the rate of five per cent, and allowing for depreciation, they were able to make a profit of $407.

Another of the experiences commented on at the conference was at Assington. This has been carefully studied by Mr. Daniel Thomson in his paper "Should Co-operative Societies Go into Farming?" printed by the Scottish Co-operative Wholesale Society. The Assington Agricultural Association, near Colchester, was founded in 1883. The two farms, Severals and Knotts, of 223 acres, were already co-operative, for they had been carried on by an association of laborers since 1829 and 1850, respectively, and were for many years a decided success—dividing 6,000 profit in the last seventeen years. Failure, however, came to the lot of the laborers, (1) because of bad seasons and low prices, and (2) insufficient capital. To these causes may be added the want of a reserve fund, which should have been built up during the years of prosperity. In 1883, the Guild of Co-operators issued a circular to co-operative societies and individuals, asking for $12,500 to start the undertaking anew on co-operative lines. A committee was appointed, and operations with the first-named farm were begun in January, 1884, when the funds collected amounted to $8,000. About half the acreage was then, and continues still to be, devoted to cereals, and the rest to root crops, pasture, etc.

Difficulties began to be experienced in 1885. There was a loss of $2,015 in 1886; and, in 1887, the committee was empowered to give up the lease if the landlord did not reduce the rent. In 1888, the land was said to be in "better heart," though a loss of over $5,000 was experienced; and the interest on loans was, in the following year, reduced from five per cent to four per cent. The trade loss in 1891 was set down at $1,750, and the total deficiency had then reached $7,770. The rent charge had been by this time reduced by $3.75 an acre—a means of relief which seems to have been felt at once though losses continued to be made in the years from 1892 to 1894. Both of the two succeeding years, 1895 and 1896, show profits of $485 and $350, respectively; and the last annual statement (1897) still shows a loss of $145. This sum, added to the previously existing deficiency, gave, January, 1897, a total indebtedness of $8,670. But the society reports at the end of 1897 a small profit and prospects of doing better this year.

This shows an ebb tide of ill fortune running for fifteen years, until it turned in 1897. But as farms all through this same period have been sought for and have been profitably cultivated in the neighborhood, agriculture cannot be said to be responsible; and as other co-operative farms have been run successfully in the same time, co-operation cannot be held guilty. It seemed to be the opinion of the members of the conference that the success at Lincoln was largely due to the fact that its farm had a direct connection with a market of its own in a large co-operative store, and that the apparent failure at Assington was due to the want of such a connection. In the latter case, too, there were intimations of faults of management. I have said "apparent" failure, because some of the students of its experience believe that at Assington the loss in the working has been made good by the new value given the land.

One of the most prosperous co-operative farms is that of the Dunfermline Society. There are 500 acres, on which it pays an average rent of $11 an acre. The capital invested is $37,500, and 1896 gave a profit of $3,125. It has been in operation for four years. The farm proper in that time has yielded a gain of $2,005, and the dairy of $11,365, a total profit of $13,370.

The worst case of all is that of the Carbrook Mains farm, of the Scottish Co-operative Wholesale Society. On its two hundred and eighty acres, employing a capital of $20,000, this society has in five years lost $26,110, or more than its total capital; but the enterprise in 1897 turned the corner and showed a profit. Its rent is now only $7 an acre, and was but $8.50 at the beginning. I got no explanation of these losses, which, however, are manifestly due to special, not general, causes; for they are entirely out of line with the average results of other co-operative farming. The profit for 1897 may be the beginning of a run of fat years.

CHAPTER III

SOME COPARTNERSHIP FARMS

The Bostal farm of the society at Woolwich, which I went to see, is run on a copartnership basis, as are all of the productive enterprises of that very flourishing association.

This Woolwich society, called the Royal Arsenal Co-operative Society, after the principal industry of the town, began in the exchange of a few ideas by some workmen at their benches, in 1868. At the first meeting, the whole amount of stock subscribed was $22.75, received in sums varying from twenty-five cents to $2.50; and the limit of stock allowed to any one member was $5.00. It took three weeks to raise a capital of $36. With this a chest of tea was bought, and a small back room, a workshop, filled with lathes and benches, taken as a store. Mr. Alex McLeod, still secretary of the society, and the proud premier of the celebration which I saw at the opening of a new branch, tells how in those early years, after doing a full day's work, he would go to London to buy the supplies for the next week's business, carry them on his back to the railway station in London, and from the railway station at Woolwich to the little store.

One evening would be spent in weighing up goods and preparing them for distribution; the shop was open for business at first on Saturday evening only. After the evening's trade, the secretary's wife would scrub the counter. The managing committee acted as both buyers and shopmen. All their services were given gratuitously; nor was it until the association had been established for nearly two years that the secretary or committee would accept any payment for their services. A business of over $2,000 was done the first year; and there was a profit for distribution of about $100. The members received a dividend of $49; $4.25 was paid for interest on capital; the stock in trade was depreciated by $38, and $6 was carried forward as a reserve. This was a good specimen of co-operative financiering. The depreciation entirely wrote out the stock in trade from the assets, and the reserve fund created was equal to five per cent of the total capital. By 1873 the sales were at the rate of $13,000 a year.

It became necessary to employ paid help; and upon the engagement of the first shopman, the right of the employé to a share of the profit was recognized. This principle has ever since been carried out by the society, and each employé who is a member receives a sum calculated at the same rate per cent on his wages as is credited to members on their purchases. This was formally put into the rules in 1877, at which time, also, a rule was adopted appropriating two and one-half per cent of the net profits of every year to the educational fund. "Never in the society's history," says *Origin and Progress*, published by it, "were resolutions arrived at that demonstrated more clearly the inner aims of the movement—education and equity."

In the panic of 1894, England was strewn with the wrecks of all sorts of land and building societies, and the town of Woolwich suffered severely. The Royal Arsenal Co-operative Society had to pay in withdrawals to its members $370,000 in two and a half years; and, thanks to the policy of accumulating reserves, so prudently inaugurated at the very beginning, it met the drain without a quiver. The society is so steadily extending its business, putting up new branches and enlarging its premises, that it now has its own building department. At the celebration I attended, these figures were given out as to its operations: Its annual sales were then $975,000; its capital $580,000; its reserve fund, $42,500; sales since the commencement, twenty-nine years ago, had been $10,000,000. Up to 1898, of the profits, $780,450 have been paid as dividends on purchases; $60,590 have been paid to employés, and $22,500 have been given to the educational department. In 1897, the sales amounted to $1,008,685, an increase of $141,125 over the preceding year; the capital had increased to $614,050, and loans to members for house-buying were $114,305, making the total $304,890.

The streets of Woolwich on the day of this demonstration were gay with flags and music, and a proud and happy crowd. A long procession was made of the wagons of the society, decorated with mottoes. One of them bore this list of the industries of the society: "Butchers, Bakers, Grocers, Drapers, Tailors, Shoemakers, Farmers, and Dairymen." Another had this: "Co-operation is the peace of industry, the opposite of competition, which is the war of industry." Another, this: "The object of co-operation is, by concert in trade, to give capital to those who have none." The great drill-hall of the town was filled in the evening with an audience of thousands of people, who had gathered to hear music and speeches by eminent co-operators.

This celebration was made the occasion of beginning the issue of *Comradeship*, which is to be the journal of the society; and a paragraph from the opening article is worth transcribing. After calling upon the co-operators to be true to the spirit of the movement in store and workshop, in federations of stores and workshops, in the widening circle of copartnership societies, which seek to capitalize the worker's gain in character and citizenship, and in the agricultural associations, by which Britain, with Ireland, is seeking to redeem the land, it says that the purpose of *Comradeship* is to make it known "that the achievement of ten thousand men and women in Woolwich is the open heritage of all; that the six stores, the finest bakery in London, the two farms, the stables for seventy horses, the sixty-five red vans, the new, healthful tailors' workshop, the boot and shoe works, the building department, the people's bank are not for individual gain, but for common good; that all are welcome to enter in, to take part in the government, to enjoy equal share in the fruits of the labor and the love of the pioneers of the past, and to earn the gratitude of the Woolwich of the future as its pioneers of the present." Any one can enter this "open heritage" by a payment of thirty-seven cents.

The results of these twenty-nine years justify this sentence from *Origin and Progress:* "The society is to-day a monument of the disinterestedness of the few workingmen founders, who, by sheer hard work, unselfishness and determination, but without any idea or hope of ultimate reward or honor, succeeded in establishing a society which has been of greater benefit to the working classes of Woolwich and Plumstead than any other agency."

This is not the boasting of him that putteth on his armor, and by no means of him that putteth it off, but of him who, having won his victory, keeps his armor to win greater ones.

Much of the land of the Bostal farm, when bought, was little better than a swamp, but I found every acre under the most smiling cultivation. The farm lies in a gently-rolling district bordering on the Thames. From its low-lying surface the river is invisible, and the ships go sailing by, plowing their furrows through the distant fields as if these, too, hastened to open themselves like the waters of the sea for the speeding of British commerce to the four quarters. The purchase of the farm which the history of the society describes as "most important" was ordered by the quarterly meeting held in May, 1886. It comprised over fifty-two acres of market garden land, with a homestead, farm buildings and cottages, the whole of which cost $31,000.

The land, when taken, was in poor condition, and great expense was incurred in bringing it into a good state of cultivation. Three old cottages were pulled down and one erected; old cowsheds were converted into piggeries, and the breeding and fattening of pigs commenced. But this last venture soon received a severe check from swine fever. In addition to this disaster, the first bailiff, who proved inefficient, had to be removed, and another, who still occupies the position, appointed in his place. During the summer of 1887, a slaughter-house was built on the farm. In the early part of 1889 new piggeries were made on the most approved sanitary principles, of brick with slate roofs and iron fittings. The yard was drained and paved, and no expense spared that would better the sanitary conditions. The result is a model piggery, which will compare favorably with any for many miles around. These additions enable the society to keep from three hundred to three hundred and fifty pigs. Further improvements were the laying of a constant supply of water to the farm buildings, provision being also made for its use in the fields for irrigation purposes; the lighting of the farm buildings and yard by gas; the covering of a good portion of the yard by a corrugated iron roof, and the erection of two greenhouses for the growing of tomatoes, cucumbers, etc.

An agitation was on foot during 1889 and 1890, for laying out the farm for building purposes. The question occupied the attention of the members at several meetings, and culminated in a special general meeting being called to consider the question on April 16, 1890, when the proposal was negatived.

The crops grown are varied, and include cabbages, colewort, greens, savoys, cauliflowers, brussels sprouts, sprouting broccoli, Scotch kale, spinach, lettuce, rhubarb, celery, onions, leeks, pickling cabbage, parsnips, turnips, carrots, beetroot, potatoes, horseradish, vegetable marrows, peas, broad beans, scarlet runners, mint, parsley, sage, and cucumbers. Mangel wurtzel, kohl-rabi, tares, vetches, and rye are also grown for the society's horses and pigs. A portion of the farm is allotted experimentally for fruit-growing. About one acre is laid down with strawberries, and about one and a half acres to various sorts of apples, pears, plums, gooseberries, raspberries, and red, white and black currants.

From the first the farm has been worked by the society, and as far as possible produce only grown that can be profitably sold or used by it. Although several causes, its report says, make the establishment, charges, and cost of labor particularly heavy, yet the working of the farm has, during a period of nearly ten years, resulted in only a slight loss on the transactions; and this (an almost nominal) loss is only arrived at after debiting the farm with five per cent interest on all capital invested. On the other hand, the farm has the advantage of a ready and certain market for its produce; does not incur the heavy charges for carriage to market and commission to market salesmen, and, to some extent, avoids the uncertainties of the different public markets.

It was a great day that I chose for my visit to Woolwich. A new branch was being opened, and the occasion was being celebrated with every festivity known to co-operators—processions, music, speeches, and, of course, a silver key to unlock the door of the new store. The manager of the farm, with the helpful courtesy to investigating strangers which is characteristic of co-operative folks, and is part of their policy, took himself away from this holiday and showed me over his scene of work. As it was in the fall, the summer glory of the crops was gone; but even in the brown decay of autumn, the farm looked a model of clean and thorough cultivation. There were still large patches of celery, and many of the plots were brilliant with the scarlet blossoms of the bean. The hothouses were full of tomato vines, running to the roof, which had yielded at the remunerative rate of $400 to the acre, but were now spent. It seemed as if not a yard of the land had been allowed to evade its tribute. Rows of cabbages were planted between the rows of celery, and from the early lettuce to the late bean, every rood was kept busy. In the extensive piggeries, floored with brick, drained, with sheltered sties for winter breeding, there was not in sight a shovelful of the wallowing mud which makes the ordinary pig-sty an abomination.

As we walked around, the manager told me his side of the story of co-operative farming. The produce of the farm was $10,000 a year— about $200 an acre. "My neighbors," he said, "can't do half that, although we got the land in bad shape." The horseradish gave $300 an acre; the beans, $200; the rhubarb, or pie-plant as we call it, $250. Some of the land, by double-cropping, yields $400 an acre. Mr. William Campbell, of Leeds, in his paper on "Co-operative Agriculture," before the Annual Congress of 1894, described the crops of vegetables as "most marvelous," and gave these figures: "In 1890 the following values at rate per acre were obtained: Beetroot $250, brussels sprouts

$195, cabbages $215, lettuce $340, onions $240, rhubarb $300, kidney beans $210, tomatoes $450, and nearly all other crops in proportion; and they not infrequently obtain a second crop."

They employ steadily about ten people, and pay higher wages than the average in the vicinity. The pay of the women, of whom there are several, was eighteen pence a day. If this was "higher than the average," what must have been the general conditions of the farm-women thereabouts? They had to have work every day to make nine shillings a week. The manager said he tried to keep indoor work enough for the bad days to save them from losing much time. The best men get $6 a week; their rent is seventy-five cents a week. No doubt, the nearness of London, whose smoke is always in full view, as well as the coöperative employer, accounts for these wages so superior to the average on the farms of England. These employés, like all who work for the Woolwich society, have a share in the profits of its operations; and they can become members of the society if they choose to do so. In some of the copartnership concerns, the employés are expected to become members. Their share of the profits is given them only in the form of credited payments on shares until they have fully paid up the minimum required. But in the Woolwich society this "compulsory capitalization," as it is sometimes called, is not practiced. The fact that the farm does not show a profit some years does not deprive the farm-workers of profits. At Woolwich, as at some other productive centers, all the profits of the various enterprises are pooled, and the employés, as well as the shareholders, participate in the general result—and this is very handsome at Woolwich.

Here, as in some other places in Co-operation land, I found the heads gave not very enthusiastic reports of the response of their employés and associates to the efforts made in their behalf. The mass, even among the members of the oldest and most successful societies, answer but sluggishly to the efforts of the leaders, as has always been the history of the mass. From the days of the Israelites, murmuring to be led back to the lashes and flesh-pots of Egypt, down to to-day, the history of emancipations has been always the same. All the present agricultural laborers on the Woolwich farm are members of the society, and get a dividend on their wages as well as on their purchases. This dividend last year (1896) was one shilling ten pence on the pound, nearly ten cents on the dollar. "They are better off than their neighbors," the manager said; but he added that there was little interest among his farm-workers in the co-operative idea and its application to themselves. But at Kettering and at other places, there was a much more encouraging atmosphere.

The manager, or "bailiff," at Woolwich, a most earnest, faithful, and hard-working man, was much mystified, not to say discouraged, by the deficits which, notwithstanding all his efforts and his apparent success, as judged by what all his neighbors were doing with their farms, were figured out by the accountants of the society. For 1896, he had it proved to him by these figures that the farm had lost $1,490. It was evident that he did not understand it, and, in fact, did not believe it. He was turning out $10,000 worth of stuff a year at a cost of only $8,000; and yet, he was told there was this big loss. I was not in a position to throw any light into his darkened mind then; but afterward, when I discovered that the farm was being charged, on the books, interest on a value of $600 an acre, I understood how artificial the deficit was. The report of the society itself speaks of its farming losses as "almost nominal."

One thing was beyond question: The results of co-operative farming here and almost everywhere else are better than the average of private enterprise. Mr. E. O. Greening pointed out, speaking in the Co-operative Congress of 1894, that the yearly loss on the farms in the United Kingdom was nearly $5 an acre, equal to the whole rent of the farms. And this manager did not have deficits always to face. For the first half of 1897 the Woolwich farm, he told me with great satisfaction, showed a profit of $375. A year or two before, a neighbor, a large landlord, had come to visit him and had boasted that that year he had cleared $5,000 on his one thousand acres. The same year the Woolwich co-operative farm had netted $50 an acre, or, as the manager proudly pointed out to me, ten times as much.

Another copartnership farm, and one of the most interesting of the experiments in co-operative agriculture, is that of the Scottish Co-operative Farming Association. It has had a long and varying career, which has just ended in liquidation, and is a capital instance of the pluck and devotion to principle which are at the bottom of the amazing pecuniary success of the co-operative movement, and give it that moral illumination which is even more important. A finer body of men than the leaders of the Scotch co-operators it would not be possible to find—shrewd and enthusiastic, full of political economy and poetry as only Scotchmen can be. Some of the best of them have been for years pioneering in the work of putting co-operative agriculture where they believe it is destined to stand—at the head of the agriculture of their country and the world. I was fortunate enough to meet one of them who has had much to do with it, Mr. James Deans, one of the lecturers and organizers of co-operation in Scotland, whose inner lineage runs back to Bobby Burns and Adam Smith and Robert Owen. As the result of several years' debate by co-operative conferences, the organization of the Scottish Co-operative Farming Association was undertaken. Appeals were made to all the co-operative societies to take stock. But capital came in very slowly.

Five years' agitation was required to raise $65,000; but this was got at last, and a constitution was drawn up and a society registered under the name given above. Then came the question of land—more difficult, even, than that of capital. Capital came from co-operators who were willing, but land had to be got from lords who were not willing. Nobody, at first, would let them have a farm. They advertised and answered advertisements; but no one had faith enough to lease land to a co-operative farming society. They were at last able to get a lease of a farm in the county of Stirling, containing 384 acres, at a rent of $7 an acre. A manager was employed, and they went to work. From the beginning, one of their principal occupations was the buying, breeding, and training of horses for the co-operative societies, which use a great many for their delivery wagons. They also began to supply milk to the Glasgow societies.

At first they did very well. The first year they paid interest on their capital and had a little left. The second year they paid six per cent on capital and a bonus of sixpence on the pound to workers and consumers on the amount of their wages and purchases, made the contribution for educational purposes, which is so dear to the co-operator, and had a little surplus besides. The third year they leased another farm in the parish of Stirling, at a rental of $11 an acre. This year they paid only interest on capital.

About this time began the disastrous fall in the prices of produce which has wrought such havoc among the farmers of Great Britain, who are tied down by their leases to heavy rents which do not fall. Hay dropped from $22.50 to $12.50 a ton; oats, potatoes, and other produce also declined severely. But in the next year they leased three more farms in the parish of

Nitshill, near Glasgow, containing 494 acres, at a rent of $13.50 an acre. They bought the best stock, and put up milk-houses and other improvements of the best pattern; for it was their only good policy, they felt, to have the best tools and the best stock. These and other improvements absorbed too much of their capital, and they went on struggling until the failure in January, 1898. Circulars and appeals have failed to induce the societies to contribute enough additional capital to work the farm to advantage, and until the last year the decline in produce has been almost uninterrupted. Other tenants had their rents reduced; but their landlords said that the co-operators were rich and could afford to pay. In spite of all these difficulties, however, they paid in 1896 five per cent on their capital, and had $2,500 of profit to credit against previous losses. The farm, in 1897, was sending from four hundred to five hundred gallons of milk every morning to the co-operative dairies of Glasgow, and was to be devoted still more largely to milk. Ten thousand acres would not be enough to supply milk to the Glasgow societies. It also supplied the great co-operative bakery, the Wholesale Society and others with hundreds of horses. As it is, the farms of the society made a better showing than the proprietary farms in the vicinity.

In January, 1898, this farming experiment was brought to an end on the application of the Scottish Co-operative Wholesale Society, to whom the farm was indebted some four thousand dollars. Its liabilities were $87,532, while its assets were $29,670 less than that. But at the same time this failure was recorded, some consolation comes from the report, as given in the *Co-operative News*, that the Carbrook Mains farm of the Scottish Wholesale, which hitherto has been the worst of all the agricultural experiments of co-operation, shows a balance on the right side this year for the first time. Among the causes of lack of success, three are very conspicuous: Want of capital, high rents, and the leases which compelled the old-fashioned rotation of crops. These obstacles are easily avoidable. Co-operators now see that they must buy land instead of leasing it; and as they have more money than they know what to do with, the supply of capital will be unlimited as soon as the co-operators have acquired a little more courage to go forward. As to rotation in crops and other oppressive conditions of leases, these will disappear as the societies become owners instead of lessees. The Scotch look with envy on the changes which have been made in the land laws of Ireland, and feel that they are far behind their Celtic brethren. "The Irish did a lot of howling," one of the Scotch co-operators said enviously, "and they reaped the benefit of it in land reforms, which have put them far ahead of us." The agricultural part of the Scottish Farming Association's experiment was successful in the highest degree. The society holds several silver cups, nine gold medals, and dozens of silver medals, gathered at the principal cattle-shows in Scotland for the superiority of its cows and horses. "We have not licked all creation," its manager said, "but we have licked a large part of Scotland."

This farm practiced profitsharing from the beginning, and allowed all its employés to become members. In fact, the rules provided that all the workers' profits were to be retained and applied to the purchase of their shares. The society maintained an educational fund, out of which it procured the agricultural magazines and sent some of its young men to attend agricultural schools. I found Mr. Deans and his associates not at all cast down by the results of their work. They look upon it as a necessary piece of pioneering. The experiment has had the effect of leading many other societies to try farming, and it has made the mistakes for others to profit by.

More societies in Scotland are every year

going into farming, in one branch or another. The Scottish Co-operative Wholesale has a farm for fattening its cattle. Most of the Scotch societies kill their own meat. The great co-operative bakery in Glasgow has already a good deal of land, and is getting more every year, for supplying itself with vegetables for its restaurants, milk, butter and hay, and for resting its horses. It has 103 acres solely for growing tea-roses and flowers for the co-operative festivals and for the private entertainments, for which it acts as caterer all through Glasgow. All the employés of the bakery share in its profits; but as the bakery is a society whose only shareholders are societies—a society of societies—the representation of the employés cannot be individual; but an organization called the Bonus Investment Company has been formed to invest the savings of employés in shares, and this society sends delegates to the meetings of the bakery society on the basis of one delegate for every one hundred and fifty members of the society, and one additional delegate for every $400 of bakery stock it holds. The laborers on the farms of the bakery have the same rights in profit and management as the other employés.

Kettering was the first place I visited after the Rugby Conference, and I found the co-operators in this center of co-operative energy full of plans for a farm they were about to start, and which, like all production at Kettering, was to be on the labor copartnership plan. The co-operative store at Kettering, which is very flourishing, and has branches scattered throughout the town, would need two thousand acres to supply its demand for milk alone. The Kettering society has been preparing for years for its agricultural venture. It has established in the country near by more than one co-operative store whose members are almost all agricultural laborers; and some of these are the best branches that it has. The Kettering society did a great deal of agitation among the farm-laborers in the vicinity, holding meetings and distributing tracts among them, in order to prepare them for this store.

At one of these meetings of the farm-laborers and others to discuss these questions, the parson sat in the body of the audience, and not in the chair, as is usually in England his prescriptive right, and an agricultural laborer sat on the platform as chairman. The parson, by the way, had not been asked, but he was interested enough to come of his own motion. The attitude of the co-operative movement towards the parson is usually one of an indifference that borders on hostility, not because he is a parson, but because of the close alliance which has usually existed between the parson and the squires, and landlords, and other powerful interests whose influence has lain across the path of the movement. An educational course of lectures proposed in Kettering last winter came to grief because the chairman of the committee thoughtlessly described the lecturer as a "co-operative parson." That word "parson" killed the whole thing. The co-operators seldom try to get the influential people in their localities to patronize their efforts. "When the big bugs take the lead," a co-operator said to me, "a society usually goes to the bad." But they often find individuals among the aristocracy who are very friendly to them. An illustrious instance of this is the present Marquis of Ripon. Lord Vernon has allowed co-operators to have land from his estates on easier terms than his other tenants. Another case is that of the owner from whom the Kettering society has obtained its land. He and his wife are very well disposed towards the co-operative movement. When I was there, the society was negotiating with their agent to get some very desirable land, and it has since closed the bargain. The members were quite unanimous and very enthusiastic about their plans, and looked forward with confidence to their ultimate success.

Having obtained the land, the next step must be to get the labor. The average wages paid about Kettering are $3.50 a week. The co-operators pay more than this, and, as one of them said to me, "ought to pay more. By paying more we get more." Only picked men will be taken. The laborers who have attended the meetings and have become members of the Tiptree store, and show that they understand the co-operative idea and are interested in it, are naturally the most progressive and active of their kind; and the offer of higher wages than are paid by the surrounding farmers will make it sure that the society gets the best men. Each one pays his shilling into the society to become a member. He then knows that the work he is doing is for his own interest, as well as for that of other members. His share of the profits accumulates until the minimum amount of shares required—which is usually $25—has become fully paid. After that he is free to let his profits accumulate, to receive interest on them as loaned capital, or to withdraw them for his own use. The search for laborers will at first be a "still hunt." Their names must be kept under cover to protect them from the farmers — now their employers. The farm which the society has secured has cottages upon it, and the laborers now renting these cottages would certainly be evicted if it were known that they were going to join the co-operative farm.

CHAPTER IV

GOLDEN VEINS OF THE EMERALD ISLE

I went over to Ireland to see the co-operative creameries—some of them copartnership, like those at Grange and Doneraile—and to look at the People's Bank at Doneraile, and gathered news there and elsewhere of a development of agricultural co-operation for which I was entirely unprepared. In the past Ireland has been the most distressing country in Europe to the tourist, with its sad faces by the roadside, farmhouses in ruins from evictions; stories on every hand of farmers, once prosperous, driven to the workhouse or insane asylum, and the indications, which are still not infrequent, that Ireland is even yet a conquered country and held by force. But, out of this desolation a new light is rising—the light of co-operation. Unexpected as it is, one now finds Ireland—the land of famines and evictions; the country which has been the most backward in Europe—further advanced in the organization of agricultural co-operation than England. This has been due to the innate intelligence of the people, roused by the efforts of a patriotic and intelligent group of men, at whose head is the Hon. Horace Plunkett, M.P., a young and able statesman of the new school—the economic school.

If one wanted to turn a phrase, he might say that Mr. Plunkett was an economic Parnell; but parallels are always misleading. Parnell devoted himself to the rescue of Ireland by political effort; Mr. Plunkett gives himself first to an economic regeneration. He does not believe in political home rule, but is ardently devoted to the idea of economic home rule. Parnell's work was largely carried on in secret; all of Mr. Plunkett's means are open to the public view. The real spring of the popular energy which vitalized Parnell's agitation was in the industrial distress of the people. They wanted relief, and their demands took a political shape because there was a political organizer of unrivaled capacity and persuasiveness, who made them believe that political remedies would set everything right. The political agitation has evaporated, but it has left behind it solid results in the reform of the land laws and the revision of rents; and without these, co-operation would certainly have been much more difficult.

Mr. Plunkett and his friends began their work by promoting the organization of co-operative creameries among the Irish farmers—a movement in which they were so successful that it soon assumed proportions too large to be carried on by merely individual effort, and they consequently formed a society—The Irish Agricultural Organization Society — under whose auspices co-operation in Ireland has been developed to a surprising prosperity. The point at which to begin co-operation was chosen with great sagacity. At a conference upon land, co-operation and the unemployed, held at Holborn town hall, in 1894, Mr. Plunkett explained why it was that the dairy was taken as the industry in which the farmer could most easily be persuaded to show what combination would effect.

Ireland, by virtue of its unrivaled soil and climate, was once the greatest butter-producing country in the world, but the application of machinery to this industry enabled countries like Denmark and Sweden to beat the Irish butter in quality and cost. Capital at once saw the advantages of applying the factory method to Ireland, and "the green pastures of the 'Golden Vein' were studded with the snow-white creameries, which announced the transfer of this great Irish industry from the tiller of the soil to the man of commerce. The newcomers were able to secure the milk at a price equivalent to its then value to the belated farmer, who deemed it advantage enough that he was saved the trouble of churning, while they (the proprietors) realized enormous profits. There was at any rate a definite, substantial object for which to combine." The point where such profits were to be saved to the farmer was the place to begin, and for five years these promoters of co-operation went about organizing the farmers to put together such money as they had, or could borrow, into co-operative creameries.

This was not an easy task. "There were immense difficulties to be overcome in inducing Irish farmers even to consider co-operative action. In the first place, volunteer association for industrial purposes was unknown in Ireland, and almost every man who prided himself on special knowledge of the Irish people, confidently declared that it was altogether alien to the national temperament and habits."* Thus, Mrs. Sidney Webb (Beatrice Potter) says in her book on the co-operative movement: " It would appear that a Celtic race is not favorable to the growth of this form of democratic self-government." The failure of co-operation to take root among the English farmers was also

* Report of the Recess Committee on the Establishment of a Department of Agriculture and Industries for Ireland. Page 360. Dublin, Browne & Nolan.

one of the deterrents, but the result has proved the accuracy of the statement made by Mr. Plunkett, that the average Irishman is more intelligent than the average Englishman; for the movement which in England so far has taken no hold on the farmers, has been a conspicuous success in Ireland.

By the end of 1893 there had been established thirty co-operative creameries. The farmers united in these creameries found that their cows returned them an increased profit estimated at from ten to thirty-five per cent. This was all the fruit of co-operation. In the words of one of Mr. Plunkett's associates: "A distinct step had been taken. It could be generally stated that a highly technical manufacture had been conducted on sound commercial principles by associations of farmers acting through committees elected under their rules from among themselves. Their product was excellent, and the venture was highly remunerative. There was not the slightest indication of even the average percentage of commercial failure being incurred. This result had been accomplished without any external aid whatsoever, except the advice and exhortation of the apostles of co-operation."*

*Report of the Recess Committee—Memorandum on Agricultural Co-operation, in Appendix N. By R. A. Anderson Secretary Irish Agricultural Organization Society.

This was the first step. Then came the second. The societies, acting singly, soon saw that they were unable to keep themselves posted with regard to the solvency of the commission men, to whom they were consigning butter in bulk, or to stay in touch with the English markets; in other words, they had no generalized commercial experience, and could not have it, as long as the movement remained in this nebulous condition. A wider organization, therefore, became necessary. A society of societies was formed to sell the product of the creameries. This was the Irish Co-operative Agency Society, which started in the autumn of 1892, with its head office in Limerick and the store in Manchester. This venture at first was not financially a success. It contracted bad debts, got into lawsuits, and lost all its capital in its first year. But the farmers persevered, and after three years of trading, the losses had been made good and the society had become a sound and apparently permanent institution. Its sales were $379,610 in 1895; by 1896 they had increased to $553,630. It is represented in every large town in England. The Irish Agricultural Wholesale Society has lately been formed to buy seeds and manures.

It was soon found that, in addition to this selling agency, a society of still broader scope was needed. This was effected by the formation, in 1894, of the Irish Agricultural Organization Society, already spoken of. The result fully justifies a statement of Mr. Plunkett that it was calculated to bring about important changes in the condition of Irish agriculture, and, we may add, changes ultimately quite as important in the social and political aspects of Irish life. During the early summer the society's plans were discussed at public meetings and in the press. The number of its adherents rapidly increased. The funds required were subscribed. A committee, drawn from both parties and all classes, as far as possible—"as strong a committee," Mr. Plunkett says, "as Irish public life could produce—was entrusted with the carrying out of the program." The new society was intended to be only temporary, being constituted for a term of five years, and its work was to be simply propagandist.

An extract from the speech of Mr. Plunkett at the inaugural meeting will show the spirit in which the society was conceived:

"The farmers, from the nature of their occupation, are incapable of evolving for themselves the principles which must be observed in framing such rules as will do justice between man and man, and harmonize the interests of all concerned. Even when a farmer grasps the idea that he ought to combine with his neighbors, he cannot put before them an intelligible and working scheme. Now, here is the point at which, without any interference with his business, without weakening his spirit of independence, without any departure from the principles of political economy, we can do the Irish farmer a great service: To bring to the help of those whose life is passed in the quiet of the field the experience which belongs to wider opportunities of observation, and a larger acquaintance with commercial and industrial affairs—that, gentlemen, is the object and aim of this society."

"The keynote" of the plan, Mr. Plunkett said in the same speech, was "that the Irish farmers must work out their own salvation;" and "this can be done only by combination among themselves." To help the people to help themselves—this is the work for the well-to-do, whom Mr. Plunkett has brought together.

Success has been beyond expectation. The number of co-operative creameries and agricultural societies has grown from thirty in 1893 to sixty-four in 1895, to seventy in 1896, and to ninety-three in 1897, and to one hundred and thirty-one in March, 1898, including branches. There are 8,750 shareholders, not including those who, without being shareholders, directly participate in the business and benefits.

The output of these societies for 1896 was $1,417,290. Almost all of these dairy societies now transact what they call agricultural business—that is, the purchase of their members' supplies for farming and the sale of their products. This, in fact, is almost always the rule with those more recently formed. The price obtained for milk has increased. Another gain in which the community-at-large is a sharer is the improvement in cleanliness and in methods. The Irish Agricultural Organization Society has succeeded in getting the commissioners of national education to direct the Dairy Instructor to consult it as to his movements, so that he can always be sent to those creameries which need his services most; and he makes his reports in duplicate to the government and the society. The latter makes it a practice to have its subsidiary societies follow out his recommendations.

One of the plans which the Organization Society has in contemplation is to select and equip a certain number of conveniently situated co-operative creameries as experimental and training centers, with experts and apparatus. Pupils can then be sent by the societies to these training schools. The course will consist of practical creamery work, simple mechanics, elementary dairy chemistry, the testing, Pasteurization and sterilization of milk, the ripening of cream, and creamery bookkeeping. There will be examinations and diplomas. The course is to be open to dairy-maids. The Organization Society now offers prizes to the creamery managers whose monthly estimates of profits come closest to the profits realized. The competition for these has been keen, and in one instance the winner of the prize had made an estimate which came within thirty-one cents of the actual result. The Donaghpatrick Co-operative Agricultural Society has taken a few acres as an experimental farm, divided it into a large number of plats, and cultivated these on the most scientific principles; cropped them with the best seed suitable to the locality, adapting the manures carefully to the crops, measured the plats, and weighed the produce. This farm is situated by the roadside. All the plats were labeled with the names of seeds and manures used in them, and the people from far and near have come to see and be instructed.

It was part of the plan, originally, that as soon as sufficient organizing help could be obtained, the co-operative method should be extended to every branch of the farmer's business. Remarkable progress has already been made in the subsidiary industries. The societies formed to buy and sell collectively farmers' products and supplies increased in number from thirty-one to forty-six during the year 1896, and now have a membership of 3,865. Mr. Horace Plunkett has shown that in one part of Ireland the farmers, by co-operation in the first year, were able to save in the cost of their materials more than the total rent paid by all the members of the association. The societies have been much more successful in their collective purchasing than in their selling, the latter being much more difficult. One of the obstacles to their success was lack of sufficient storage room to hold grain for a better market. This, in some countries, is provided by the state, as in Russia, where the farmers are given the use of extensive granaries, receiving an advance from the state of two-thirds or three-quarters of the value of the grain, to enable the farmer to meet his most pressing obligations until the moment arrives for selling. In the sale of cattle and sheep, the societies have been more successful than in handling grain. The prospects of this business are considered encouraging. The farmers will obtain higher prices for their live stock, will save freights by consigning larger quantities, and the consumer of meat and the cattle in transit will gain by the adoption of humaner methods of transportation.

One remarkable feat achieved by these societies, as purchasers, was the smashing of the

Manure Manufacturers' Alliance, a ring of dealers in chemical manures. The alliance was broken by offering some of its members an amount of trade which they could not resist, and the rest of them found it impossible to keep up the combination. Chemical manures are now being bought at far lower prices than they have ever previously sold for in Ireland, and the whole country gets the benefit of the reduction which was forced by these co-operative efforts. The prices of farm seed and feed stuffs have also been reduced. Some of the manufacturers of agricultural machinery refused to deal with the societies on account of their prejudice against co-operation, but they are now coming to see that these are permanent and that their trade is worth having. Some of the foremost of them now give the co-operators special terms.

Of quite as much importance as the reduction in price has been the gain in purity. The consumer now no longer has to accept the statement of the manufacturer as to the quality of manures, but always buys them subject to analysis. The saving on manures, seeds, and other agricultural requirements is estimated to have amounted to not less than $1,250,000. The profits of the creameries have been $906,765. These give a total of $2,156,765. The expenses of organizing the movement since its inception in 1889 amount to about $40,000. Thus it follows that, taking the seven years, every $1,000 spent on this work has produced over $50,000 of actual cash benefit to Irish agriculturists.

The English spend $25,000,000 a year for eggs. Ireland supplies a great many of them; but, owing to ignorance in breeding and carelessness in marketing eggs of all ages together, the Irish egg has fallen into disrepute. The Organization Society has employed a poultry expert to travel about from society to society, giving instructions with regard to the best class of fowls to breed, and how to market the eggs in fresh and attractive shape. In this way the egg producer will be able to obtain for Irish eggs, as the co-operative dairies have done for Irish butter, the highest market price instead of the lowest.

There are now thirteen people's banks in Ireland under the auspices of the society. This is only a beginning of the task it has assumed of delivering the farmers from the usurer and the "Gombeen man" by the organization of agricultural credit, from which such benefits have accrued to the farmers of foreign countries, especially Germany and Italy.

In looking over the books of the Doneraile Bank, I saw that the applications for loans were, without exception, from men of the slightest means and for the humblest purposes—the purchase of a calf or pig, or seeds, or a cart-horse, or funds to take a small contract. The accounts show that the interest and principal were paid with scrupulous punctuality. It is an axiom with bankers that no borrowers are so honest and prompt as the smallest and poorest, and no banking so safe as co-operative banking. The results fully justified the remarks of the auditor: "Your bank has made its way among the people without solicitation or advertisement. .... It can now be taken as an example over the country. It illustrates the unspeakable benefits which may be conferred by co-operative banks." The secretary of the Irish Agricultural Organization Society adds to this that, though the transactions of the Doneraile Bank were most humble, "they were of the greatest moment to those small farmers and farm-laborers, hitherto without credit except at usurious rates and on demoralizing terms, to whom new credit facilities were thus afforded. Nothing has been more encouraging to us than the way in which the wealthier farmers have attended to the affairs of this little enterprise for the sake of its financial and moral advantages to their poorer neighbors."

In an article in a volume published by the *Irish Homestead*, the organ of the Irish Agricultural Organization Society, at the time of the Irish Textile Exposition, Mr. Plunkett pointed out why co-operation had to begin in Ireland with agriculture. There is no factory population in Ireland, and the building up of an industrial community, therefore, had to commence with the people on the land. Beginning with agriculture, also, it gave an ample and immediate return in money. But, as we have said, the plans of these Irish co-operators look far beyond agriculture. They have established a number of pioneer co-operative industrial societies. One of these was started at Dalkey, near Dublin, and is operated on the labor copartnership basis. We give this outline of the constitution of the society:

"Workers are not required, on entering the society, to pay anything, but they are all obliged to become shareholders. This they can do by allowing their share of the profits to be devoted to the purchase of their shares till they are fully paid up. This does not, however, diminish in any way their ordinary wages. The workers are paid according to the amount and quality of their work. The profits of the society, as ascertained when the accounts are made up at the end of each half year, are divided among the workers in proportion to the wages that each has earned during the time. A committee is elected by the members of the society, and the rules for hours of labor and the general conduct of business are made by the committee. A member cannot be dismissed for any cause whatsoever, except by a vote of the whole society Before a worker is admitted a member, she must first enter the work-room as an apprentice, or as a paid hand. If she does not prove herself capable and industrious, she will not be admitted into the society."

The work done by this society is principally embroidery, and its product holds its own with the best embroidery produced on the continent. All the workers are taught drawing. "Many a toiler," says Mr. J. K. Montgomery, in *Labor Copartnership* for December, 1897, who supplies the information just given, "in the smoky cities of England would envy the pleasant circumstances under which these Dalkey co-operators work; for the bright and airy rooms in which they sit command a charming view over the broad bay of Dublin."

As the increase of the creameries, to some extent, takes away employment from farmers' wives and daughters, the Organization Society intends to push this work of establishing home-industries societies, as far as possible, in every district where a creamery or agricultural society exists. Wicker work, basket making, wood carving, lace making, poultry raising, and hand-loom weaving, are among the industries to be promoted in this way. Pig-feeders' societies have also been formed, and the price of pigs has always advanced in their neighborhood. Important work is being done in stimulating the flax-growing industry, and in the selling of members' wool directly to manufacturers. The Ballinagore Society, by bulking its members' wool and having it manufactured into tweeds for them, was able to get fully two cents a pound more than the highest market price for the year.

The Board of Works of the Irish government has, by statute, the power to make loans to individuals for industrial uses, and the Organization Society has succeeded in inducing the board to consent to make such loans to the co-operative societies upon the same terms as to individuals. These loans are wanted for the erection of creameries, flax-mills, storehouses, etc. Very little has yet been done under this head, as the requirements of the Board of Works, with regard to security, are very severe. But it is believed that these will be modified, and if so, the facilities that would then be offered would be of great advantage.

The number of societies affiliated with the Organization Society increased during 1896 from twenty-nine to sixty-one, and the general meeting, which is held annually, of all the societies, under the auspices of the Organization Society, is becoming an important gathering. It is a central body for intelligence and organization, representing all that is progressive in the agricultural life of the island. While the movement to give Ireland a political parliament lags, the co-operators have succeeded in establishing this economic parliament, whose debates and reports must be of incalculable usefulness to those who take part in them; and the utterances of this body with regard to measures, political and otherwise, which are required for the promotion of Irish prosperity, cannot fail to command most careful consideration from the government.

No one can visit these creameries and the offices of the Irish Agricultural Wholesale Society, or see and hear Mr. Plunkett and Mr. Anderson, the president and secretary of the Irish Agricultural Organization Society at the conferences of farmers, as I saw and heard them, for instance, at Thurles, and not be deeply impressed by the work the farmers and these friends of theirs are carrying out, and with the evidences of its success. The principle of labor copartnership, though not recognized in all or a majority of the enterprises, has a firm foothold, as at Grange and Doneraile, and at other places.

Wages of farm laborers in Ireland are lower than even the pitiful level of England. I found farm laborers receiving $1.25 a week, with their board, if unmarried. If married, they get the free use of a house and half an acre of land, and $1.87 a week "the year round," wet and dry, with allowance of one pint of milk a day. An interesting thing is being done by the boards of guardians in various localities in Ireland for housing the laborers. They are building cottages for them, as in the vicinity of Limerick, at a cost of $550 for the house and $225 for the half acre of land that goes with it—a total of $775. This is then rented to the laborer for twenty-five or thirty cents a week. This leaves a large deficit on the investment, which is met out of the rates. It had much the appearance of an attempt to keep labor cheap for the land-holders at the cost of the general taxpayer.

When I was there, the Grange creamery was disturbed by the invasion of its territory by the English Co-operative Wholesale Society. This is building a large creamery within three miles of the Grange society, to be run upon a capitalistic basis, while the Grange gives a dividend to its laborers. The Scottish Co-operative Wholesale Society of Glasgow has recently followed this unco-operative example by establishing a creamery at Inniskillen. It has been recently stated that at present the English Wholesale has fourteen main creameries and seven branches in Ireland, and that in a short time it will, it hopes, have one hundred. Like it, the Scottish Wholesale society is preparing to make its invasion of the field of its Irish fellow co-operators more extensive.

The men who are doing so much to lift the Irish people into the higher life of co-operation look with dismay upon these tactics, which they consider in direct violation of the principles of the movement. It was co-operation which gave the workingmen of Scotland and England the money to come over to Ireland to buy Irish butter. "When they turn that capital into a means of exploiting the Irish farmer," says the *Irish Homestead*, "they are false to the principle which has brought such enormous benefits to themselves. The very weapon which they found so efficient to defend themselves against being exploited as consumers they are now converting into a weapon for exploiting the producers." But the *Irish Homestead* notes with satisfaction that wherever the Irish Agricultural Organization Society has organized the farmers, these have been able to rout the Wholesale. The annual conference of delegates from the co-operative dairy and agricultural societies in Ireland, in November, 1896, passed this resolution on this question: "This conference hereby condemns the action of the Co-operative Wholesale Society in erecting creameries throughout the south of Ireland as injurious to the farmer's interest individually, and prejudicial to the growth of the co-operative movement through Ireland."

The promotion of co-operation by individual effort was the first step; the formation of the Irish Agricultural Organization Society, to give a more effective guidance, was the second step in a progressive program in which the third step, not less important than the other two, has been taken by Mr. Plunkett, in a farsighted effort to procure from the British government the creation of a department of agriculture and industries for Ireland. In 1895 Mr. Plunkett issued an invitation to a large number of influential Irishmen of all parties to take part in the formation of a committee which should consider matters affecting the social and material interests of the country. He very carefully guarded his proposals from the slightest aspect of political design; asking all the representatives of Ireland in Parliament and other prominent Irishmen to unite with him in discussing measures for the material and social advancement of the country upon which they could unite, irrespective of their political differences. The response with which he met was most gratifying. He achieved the task, not always easy, of getting Irishmen of different parties to unite. The Dublin municipal council unanimously adopted resolutions approving of the formation of the committee suggested, and gave rooms in the Mansion House for the use of the conference.

A committee, called the Recess Committee, was appointed to report on the measures taken to promote the interests of agriculture by all the leading governments of Europe. It found that it could not gather from existing documents the information desired, and, therefore, made direct and independent inquiries for itself. Special commissioners were sent to France, Belgium, Holland, Denmark, Bavaria, Würtemberg, Austria, Hungary, and Switzerland. M. Tisserand, probably the greatest living authority on state aid to agriculture, furnished a valuable contribution, which had great effect on the conclusions of the committee. The report embodying the results of these investigations is so valuable that a large part of it has been copied by the United States Agricultural Department in its Year Book for 1897. It reviews the present economic condition of Ireland, and points out the various measures which would stimulate an industrial revival in the country, following the experience of the other European nations, and then suggests to the government the creation of the administrative machinery by which these measures could best be put into effect—a department of agriculture and industries, with a minister at its head, who should be, like the other ministers of the government, responsible to Parliament. This was not proposed in any spirit of deviation from the fundamental principle of the co-operative movement in Ireland, that the economic salvation of the people must be mainly brought about by their own efforts, but aimed only to obtain from the government the guaranty to the Irish people of a free opportunity to help themselves.

"The state steps in, abroad," Mr. Plunkett said in his address at the second annual conference of the Organization Society, "to do certain things for our competitors which they cannot do for themselves; and the state must be asked to place us upon an equality in this respect." This equalization of opportunity the Irish did not ask for in the shape of any bounties or subsidies, or high-tariff protection. They did not ask the government to do for them anything like the creamery socialism on which the Canadian government has embarked for its farmers. There the government is erecting creameries, taking the milk from the people, making it into butter and cheese, putting these upon the London market, and merely charging the farmers a certain amount per pound for the freight. The Irish farmer has learned how to do these things for himself. All that the state was asked to do for the Irish farmer by this conference was to help the restoration of agriculture and industry, by giving the people a system of adequate and practical education; by helping them to organize for themselves; by furnishing, through regular government reports, expert information; by assisting in the introduction of new industries, and by according the farmers and the industrial classes an organized representation and a voice in the administration of a department which should have this policy in charge. The government recognized the importance of the proposals of the conference by introducing a bill to establish the board of agriculture and industries suggested for Ireland. The bill was not all that was desired, but it was welcomed by the co-operative societies as a great advance in the attitude of the British government towards Ireland, and as a nucleus of what should prove a most important means in assisting the Irish producer in his competition with the foreigner and in the development of the resources of the country.

Such is a rapid review of a work which, considering the peculiarly depressed condition of Irish agriculture, the suppressive policy which has been pursued towards Irish industries by the British government for centuries, and the effects produced thereby upon the Irish mind, as well as the Irish body, must be regarded as one of the most interesting experiments in the recent history of social reconstruction. The attempt to introduce co-operation in this country, among these people, was decried as a wild experiment; but it was with justifiable elation that Mr. Plunkett, at the last annual general meeting in May, 1897, gave this summary of the eight-years' fruits of a work which has now extended to almost every county in Ireland. "Your committee," he said, "have to report that they find the Irish farmers enterprising and receptive of new ideas, self-reliant, cautious, business-like; in a word, possessed of all the qualities which entitle them to the utmost confidence in all that relates to the development of their industry." The report of the eight-years' work was, he said, "the history of associations composed almost exclusively of working farmers engaged in the transaction of the daily business of their members—in improving the conditions of their industry—applying to its development the most modern and complex system of commercial organization, with its federation acting as agencies for the sale and distribution of the produce of Irish farms in the large cities of England, including London, where they have to meet the competition of the world. In eight years these societies of frieze-coated farmers, in spite of their small beginnings, have transacted business amounting to close on a million sterling. And I think it would be easy to demonstrate that they have saved over a quarter of a million of profit for themselves."

There is nothing like this in England. There are a great many co-operative creameries in America, but the Irish farmers have not only formed themselves into individual societies, but have connected these societies into a central organization. They have even compelled to listen to their suggestion for an agricultural department the government which is not usually very ready to listen to any demands for reform from its colonies or dependencies, and which would listen to Ireland last of all. These Irish societies, moreover, are about to federate and compel the railroads to abate their extortions. Meanwhile, the English farmer has done, as yet, practically nothing. Co-operative agriculture in England has made a fine start, but it has not been in any case done by the farmers. Why the English farmer has been so backward is, perhaps, due in part to the absence of a leader like Mr. Plunkett—at one of the agricultural conferences in England it was suggested that England needed an English Agricultural Organization society,—and partly to characteristics which may be indicated by such incidents as follow.

In the discussion of co-operative agriculture before the Congress of 1896, Mr. Thomas Blandford described how the chairman of a railway company had recently told a meeting of his stockholders of the efforts that had been made to get the tenant farmers to co-operate with the railway company in the marketing of their produce. It had offered to collect their stuff for nothing and distribute it for them to the consumer, but so far it had not been able to make any arrangement with any of them. They suspected each other. Each was unwilling the others should have an opportunity of learning where and at what price his produce was sold. The Lincoln society desirous that some of its branches, whose members include every grade of agricultural worker, should imitate its success in co-operative farming, brought the matter up at a conference of delegates from these branches. These branches had accumulated $71,000 of capital; they had a market at their own doors for a portion of their produce, and a bank from which the agricultural laborer, who has become or wishes to become a small producer, can borrow at easy rates of repayment. The Lincoln society suggested to the delegates that, with all these advantages and the benefit of constant contact and interchange of ideas with the co-operators of the town, they should make a forward movement and attempt co-operative farming. The delegates made replies, which have been summarized as follows: "As a class, we still lack confidence in each other; we have no sufficiently free access to the land; we are slow of being convinced that technical education is needful for us in dairying or any other branch. We want town co-operators to take the lead and bring us in, as has been done in co-operative shopkeeping." If there was a similar spirit among the Irish farmers, it has been overcome by skillful and inspiring leadership.

CHAPTER V

GROUND RENTS OF BROTHERHOOD

There are revenues from co-operative farms, and rich ones, which do not show in the profit-and-loss account. Mr. George Jacob Holyoake, the Nestor of English co-operation, described to me, with enthusiasm, how he had seen, at Newcastle, two thousand co-operators picnicking on a sloping hill on their own land. "Slaves unchained," he said they were. It is estimated that during the past thirty years over 2,500,000 of men have migrated from the rural districts of England to the large towns; and during that time the agricultural lands of Great Britain have declined $500,000,000 in value. In seventy-five years 27,500,000 acres have gone out from under the plow. The agricultural laborers, unable to find work in the country, have had to pour into the towns, where they increased the competition for employment. The Earl of Winchilsea, in his address before the Co-operative Congress at Woolwich, in 1896, stated that the great coal lockout of 1893 was mainly due to the immense number of agricultural laborers who crowded into the mining districts and competed with the miners in their work, underselling them, and creating the conditions which led to that great commercial war. Every co-operative farm that succeeds and encourages the movement of co-operative capital and energy towards the land, arrests this depopulation of the country districts and the congestion of the centers of industry, removes one great cause for the decline of wages, and lessens the burden of pauperism on the taxes—and these taxes fall heaviest upon the working classes.

The picnic as to which Mr. Holyoake felt so much pride is but one intimation of many benefits that will flow from the possession of land by co-operative societies. The co-operators are already looking forward, as their discussions of this question show, to the time when many manufacturing industries, which could easily be carried on in the country, in the midst of its cheerful surroundings and its superior conditions, shall be transferred thither. "Imagine for a moment," said Mr. Wm. Campbell, speaking before the Sunderland Congress of 1894, "the happiness that would be created, the pleasure, comfort, and enjoyment realized by taking the poor sempstress and the sweated shirtmaker out of the immoral atmosphere of the slums and the horrid dens in which many of them have to live and toil, into the pleasant sunshine and pure air of the country. There would be times when a large amount of extra unskilled labor of a light character would be required in gathering fruit and other crops; thousands of pounds are lost annually in districts where this is not available. By the scheme we suggest, it would be always at command; and what a delightful and agreeable change to the workers at sedentary employments would be a few hours' labor now and then in the open air!"

An indispensable part of the program of co-operative farming is the education and elevation of the agricultural laborer. The benefits that will accrue from this to the co-operators and the laborers are only a part of the social wealth that will be created. In the address I heard him give before the Rugby Conference, Mr. Henry Vivian, of the Labor Association, laid special emphasis on this culture of the agricultural laborer. "Meetings of the agricultural laborers," he said, "should be convened in the districts and proposals for action placed before them, and they should be asked to provide some of the capital required. No start should be made unless the laborers show an active interest in the project, and we must keep preaching until we secure their active co-operation. This is essential, to my mind. Our work is to educate people to want co-operative workshops and employment, not to impose them on people that do n't want them."

Kettering co-operators, as already stated, have proceeded along these lines in their farming project. It is uniformly found that the agricultural laborers are very eager to profit by their opportunities to learn. Mr. McInnes, of the very successful Lincoln society, said at the Congress of 1897, that the agricultural laborers are docile and anxious to be taught the higher forms of co-operation. "With education, the peasants will be able to hew out farms from the wilderness at home just as well as in the prairie and the forest. The co-operative societies," he continued, "are able from their profits and surplus capital to assist the agricultural laborers in working out their own salvation." A delegate at a co-operative congress asked: "Why does not farming pay? The farmer said it was because the labor was too dear. The laborer said it was because it was too cheap. I agree with the laborer. Skilled labor cannot be got except at a good price; and the present low condition of farming is largely due to the degraded condition of the laborer." How much this education of the agricultural laborer may mean to the country can easily be appreciated from the very striking testimonial given in an address before the Woolwich Co-operative Congress of 1896, by the Earl of Winchilsea, to the business capacity shown by the workingmen in the co-operative movement. He said: "It is astonishing to me, as a man of business, how these great businesses can be worked and administered as they are by workingmen. It is a proof, I think, of the advance we have made during the lifetime of the present generation. The probability is that no such development and no such management would have been possible sixty, seventy, or eighty years ago. But the co-operative movement is more than that. It is not only a proof that these results have been attained, but I view it as the greatest possible educational factor in order to arrive at results still greater."

When land has been misused and becomes exhausted, the English farmer says it "has no heart;" and he describes the process of bringing it up again into condition as giving it "heart." The agricultural laborer, too, has been abused and exhausted. The copartnership plan of co-operative farming will certainly give "heart" to the laborer as well as to the land. The most energetic and enterprising, and certainly not the least successful, of the co-operators who are devoting themselves to the solution of this problem of co-operation on the land, are devoted to copartnership. It was an essential part of the plan of operations sketched by Mr. Vivian at the Rugby conference. Success was not likely, Mr. Vivian told the conference, unless to all their other resources of capital and markets the co-operators were able to add the enlistment of the active interest of the agricultural laborers; and this was to be achieved, not by making him a present of the garment of co-operative prosperity readymade, but by rousing him through education so thoroughly that he would be willing to find a portion of the capital needed, even though only a slight one, and to accept his share of responsibility for the profitableness of the undertaking.

A practical benefit that would be thus secured, was pointed out by Mr. Ritchie, of Edinburg, at the Congress of 1887. An enormous leakage, he said, occurred in farming owing to the bad condition of the labor question. In harvest time, work lagged and hundreds of pounds were lost. The success of co-operative farming would depend on giving the laborers an interest in the profits. If co-operators engaged in farming did not share profits with the laborer, more would be lost in leakage than would be gained by withholding those profits. This is confirmed by Mr. E. O. Greening. "In England," Mr. Greening said in the Co-operative Congress of 1894, "five laborers are needed to do the work that two men will do in Canada and the United States. Efficient work could only be got by giving the workers a share in the profits." Mr. William Campbell, in the paper already quoted from, says: "Every one employed on the land should feel an interest in his work, and be induced to give his best energies in prosecuting it. This can be secured only by giving every worker the right to participate in the results of his labor."

The remarkable case of Ralahine is an almost classical instance, frequently cited by co-operators, of what can be done with an apparently almost hopeless human material by treating it with justice and letting it know that it is to share in the fruits of its own industry. The story is told in full in the "History of Ralahine and Co-operative Farming," by E. T. Craig, published by Trübner & Co., of London. Ralahine was a farm in the midst of one of the most turbulent and dangerous "White Boy" districts in Ireland, where the people were ragged, hungry, and lawless, and where the lives of landlords and stewards were in deadly peril. After a bullet had been sent through the skull of his steward, the owner of Ralahine fled in 1831, leaving his house in charge of the armed police, and sought safety in the city of Limerick. He had long been planning a co-operative farm as a remedy for the misery of the people, and this crisis precipitated action. He had the luck to find just the right kind of practical enthusiast to make his novel scheme a success. This was an Englishman, Mr. E. T. Craig, the author of the "History of Ralahine." It can be imagined how much the people knew of the principles of co-operation; but when Mr. Craig explained to the laborers that they were to be their own masters; that the work of the farm was to be apportioned by a committee of nine members chosen by themselves; that there was to be no steward, but that every one should do his part of the actual work, and that they were all to share in the produce of the farm, the very ringleaders of the previous disorder became the best workers.

An almost communal system of life was adopted; the cooking was in common; the people went into associated homes. They became anxious about the education of their children; they worked well and successfully. The farm was improved, and new dwellings were erected. A farm store was opened for the supply of goods, and on account of the scarcity of money, Labor Notes were printed and issued. The people took these willingly in payment for their labor, because they found them "as good as gold" at the store.*

In the short three years that the experiment lasted, the people became wonderfully changed. They left off drinking; they kept their homes clean; they paid their rent; disorder and violence ceased; intemperance became almost unknown; the laborers, men and women alike, had earned as much as was paid by neighboring farmers, and a fifth or a fourth more. The experiment came to a sud-

*"The notes were printed on stiff cardboard, about the size of a gentleman's ordinary address card, being two-and-a-half inches long by one-and-a-half wide. The unit note represented the value of one day's labour, thus:—

---

8 PENCE

FOR ONE DAY'S LABOUR

"There were other notes representing a half, a quarter, an eighth, and a sixteenth of a day's labour, respectively, or the equivalents for fourpence, twopence, one penny, and a halfpenny. The usual wages of a labourer in the district at the time was the foundation of the rates affixed as value. Above the unit notes were others, representing one shilling, or one day and a half's labour, printed in red ink for distinction, and another equal to three days' labour, or two shillings. These notes were received as currency at the store for articles produced on the farm, for materials for clothing, fuel, rent, etc. The labour was recorded daily on a 'labour sheet,' which was exposed to view during the following week. The members could work or not, at their own discretion. If no work no record, and therefore no pay. Practically, the arrangement was of great use. There were no idlers, and as the currency was only available within the community, the members gradually found themselves in possession of ample funds after their current expenses were discharged.

"At the first issue of the labour notes there was some objection to them on the ground that 'they were not money.' But when it was found that on application to me they could be exchanged for hard coin when necessary, or for food, clothing, etc., at their own store, the objection died away, and ultimately the notes were preferred to cash payments."—*History of Ralahine*, p. 75.

den end on account of the bankruptcy of the proprietor, which brought in new owners who had no sympathy with what was being done; but the incident remains a splendid illustration of what can be accomplished with men and land under apparently the most difficult conditions, if the principle of brotherhood be appealed to.

"The results, however," says Mr. Craig, "fully justified the experiment. The landlord was relieved from a vast amount of care and anxiety, and the people were industrious, contented, and happy. Many of the peasantry began to hope that other large landed proprietors would adopt the system. On one occasion, when returning from Limerick, a Catholic priest met me, and insisted on my going to his house and partaking of refreshment, when he spoke in warm commendation of what I had accomplished, and expressed his high gratification at the change and improvement among the people at Ralahine. A few months previously the priests and the clergy were utterly powerless to quell the riotous proceedings of the peasantry, who were rendered furious by their wretchedness. The transformation of the conditions seemed a mystery to many. Social science was more successful than military force. Socialism could induce poor, ignorant Irish peasants to live in peace and harmony with each other. In their individualism, every man was for himself at the expense of every other person. Now, each found his interest promoted by promoting the comfort and happiness of all around him."

When the Rochdale pioneers came to the English workmen of 1844, and Mr. Plunkett and the Irish Agricultural Organization Society sought to rescue the Irish peasant of 1893, they found but a narrow margin left between the people and absolute extinction. But by this narrow pass English workmen and Irish farmers have escaped to a prosperity which, however incomplete it is still, would have seemed impossible when they began. The constitutional abolition of industrial injustice for the whole nation must be a slower process, but it will be all the surer at the hands of those who have disciplined themselves into such self-help and success. Those who have been able to combine economically will be all the more able to combine politically. The men who have tasted the sweet fruits of "each for all and all for each," and of equity and fraternity in a small field, will not rest content until they have seen the same principles applied progressively to every field. The men who, inch by inch, have pulled themselves out of the slough, will never be pushed back into it alive—neither English workmen nor Irish farmers. These co-operators will be the backbone of land reform, when it comes, and of other reforms.

There are the best of reasons to believe that co-operative farming has in it capabilities which promise more success than individual farming. The results of co-operative farming thus far have been very much more favorable than those of the average farming of Great Britain. Farming is a very uncertain occupation, making a large profit one year and a large loss the next. It can, therefore, be best carried on under a system which accumulates reserves in the fat years to meet the shortages of the lean years. Co-operative bookkeeping carries every year a portion of the profits to a reserve fund to meet reverses when they come. The amount of capital which co-operative farming, when it wins the favor of the co-operative societies, can command, makes it the most powerful of all the practical agencies now at work to settle the land question in Great Britain. Mr. Wm. Campbell, before the Sunderland Congress, speaks of "the almost unlimited power of raising capital with little effort or self-sacrifice." Mr. H. Vivian, at Rugby, refers to "the enormous surplus capital" and "large organized market" which co-operation offers. "Most societies have sufficient capital," says Mr. J. J. Cain, speaking of co-operative farming at the North Lonsdale Conference, "in fact, with many, money is a positive drug." The co-operative societies are more and more going into a variety of industries; and the practice, as at Woolwich, Glasgow and elsewhere, of pooling the results of all the industries of a society into a common fund before it is divided, will give co-operative farming a strength that farming dependent upon itself alone could never acquire.

A very practical matter which will help co-operators to succeed where others will fail is the necessity, which is increasing in the modern market, of uniformity of quality and supply. This could not be better stated than by Lord Winchilsea, at the Woolwich congress. He pointed out that the English were losing the command of their own markets to the foreigners because the latter organized their production and means of distribution; and he chose butter-making as an instance: "Now, the making of butter, in order to sell it according to the conditions required by modern circumstances, is a thing that cannot be carried on by individuals, but it must be done on the principle of combination. The reason of this is very simple, and even those who may not be acquainted with agriculture will see the point when I state that modern trade requires immense quantities of one article, uniform in quality and appearance, regularly delivered every day all the year round. When the Co-operative Wholesale societies go into the market to buy for your stores, that is what they desire to have; you want to have the stuff up to sample, and to be assured that it will always be the same when you buy it. But it is impossible, under the present conditions in which butter is manufactured, to secure anything of the kind. You may go to a thousand different farmers, all producing butter—some good, some not very good—but you will not find, out of those thousand samples, any that is exactly like the others. The whole of that production is, therefore, cut out more and more from our own markets by the productions of countries where the farmers have combined to produce their butter under the best conditions."

The Irish farmer, quicker than the English farmer, has seen this; and, as we have shown above, is already far along on the road of success in this organization of agriculture. What is true of dairy products is coming to be more and more true of all the products of the farm. That agricultural interest which has at its command unlimited capital and markets already fully organized, which can bring produce of uniform grade and regular supply to the consumer, and can get from the laborer his highest working power, must certainly, in the long run, "compete out" the private farmer and the individual capitalist. The amount of land that the co-operators could use is almost without limit. Hundreds of thousands of acres would be required for milk alone by the present co-operative population; hundreds of thousands more for the butter; one million acres for the meat. Then there are vegetables, fruit, and grain. "There is no reason," said Mr. Campbell, "why every large and wealthy society should not have its industrial village on its own estate within the next twenty years." The whole trend of co-operative discussion points to a steady movement towards the consummation of that possibility. I have never seen a more alert and attentive audience than that which, at the Rugby conference, took part in the discussion of this subject. It was not a lecture audience. It was a meeting of business men and women, considering a matter of absorbing interest to themselves; and there was in it, besides, the fire of determination and enthusiasm of men animated by the belief that they are engaged in a great cause. As one of the speakers in the conference said,

"Business success is not enough. We must have intellectual and moral profits as well."

The failures of co-operative farming, so far, cannot be charged to co-operation, because some of the co-operative societies have been successful. They cannot be attributed to any inherent unprofitableness in agriculture, because, within easy reach of the failures, other farms have been operated with conspicuous success. Besides the mistakes of the co-operators, which time and experience will take care of, the chief difficulties have been high rents; oppressive leases; exhausted land; extortionate rates charged by the railroads. It has been shown by parliamentary returns that the farmers in England have paid $14 an acre a year in rent, taxes and poor rates for land on which to grow wheat. The landlords and their rents and leases—one case is mentioned of a lease by which not a single straw could be sold from the farm, but all must be converted into manure, though straw is worth $15 a ton and fertilizers can be bought at $5 —can be got rid of, as they are being got rid of, by purchase. The railroads are coming to see that they are only killing themselves by killing the industry along their lines; and the very roads which have been charging three times as much for carrying produce from Kent to London as was charged for bringing it all the way from Belgium to London, are now asking of the co-operators what they ought to do to encourage the industry along their lines. "If you will tell us what to do, we will do it," they say. In giving "heart" to bad land and poor labor, and establishing quick and certain connections with the markets, none can have resources equal to the co-operator.

Co-operative agriculture is well started in England, Scotland, and Ireland. In England, townspeople through the stores are the prime movers; in Ireland, it is the farmer—a working peasant. In both, the dairy seems to be the branch of farming which first invites co-operative enterprise. All through the continent of Europe, agriculture is passing into the co-operative phase; and the evidence which we have cited seems to show that Great Britain and Ireland are entering upon the same course. These efforts of the co-operative workingmen seem infinitely vigorous in comparison with such a philanthropy as that in which the poet Tennyson engaged. As told in his life by his son, he joined with others in an attempt to settle agricultural laborers on the land, and the experiment ended by the deportation of a number of farm laborers to a little colony called Tennyson, in South Africa. The co-operative movement keeps the bone and sinew of England where it belongs, on England's land.

Every day there are more co-operators who see that the great stroke of co-operative policy now is to get on to the land. Land is cheap, and there is plenty of it to be had. With the possession of land, the circle of co-operation will be completed. With stores, factories, farms, water-power, minerals, co-operators will be not only "a state within a state," as they have been called, but in the event of decadence of the old order would really be the only state. Theirs only would be the organization whose members were held together on the basis of that reciprocity which is the true equality. Depression of business, attacks of private traders, social disorders might beat then upon the co-operative house, but it would not fall, because it was founded on a rock—on something better than a rock as a foundation for a community—dirt.

I found no subject so certain to command the interested attention of English co-operators as that of village communities, and it is a safe prediction that experiments in that direction will multiply in the near future. In all this we see co-operation coming around to its first idea. It was part of the original plan of the Rochdale pioneers, as set forth by them, "to establish a self-supporting home colony of united interests, and to assist other societies to establish such colonies." It is not too much to say that the community idea was always the ruling thought in the co-operative movement. It appeared in plans formed by men like James Hole, Lloyd Jones, and others prominent in the foundation of the Leeds society. They aimed at "the self-employment of members in co-operative workshops, and finally the collective organization of an industrial city, self-supported, self-sustained, self-directed, for the benefit of the whole."* Robert Owen made no mistake when he conceived of the co-operative community as the highest realization of industrial reform; but it was an error of enthusiasm for him to suppose that his movement could begin with the perfection with which it is, we may hope, destined to end.

My observations were too cursory to entitle the results to be called conclusions; but the sum of my impressions with regard to co-operative agriculture in Great Britain, may be stated as follows:

First. It has made a beginning full of promise, showing at the very start a greater success than the average of private farming.

Second. This is but the beginning.

* George Jacob Holyoake's History of the Leeds Society, p. 51.

Third. It has all the resources behind it, financial and other, needed for the accomplishment of all it may undertake.

Fourth. It is animated by the conscious hope of helping in the solution of the land question for the whole people.

Fifth. It is the most powerful of all the interests, reformatory or practical, that have set themselves at work on this, the question of questions.

Sixth. It is likely, in the next two generations, to produce results, commercial, social, and moral, not less momentous than those wrought in trade and industry by co-operation in the last two generations.

It is not farmers nor farm laborers who in England and Scotland are organizing the co-operative farm. They show little of the mutual spirit which in Ireland and on the Continent is remaking country life. British co-operative farms are the ventures of the co-operative store as this is the enterprise not of traders but of the workingman. Some of the most promising efforts are in the hands of societies like those at Woolwich, Glasgow, and Kettering, where all co-workers can share in ownership, and we may hope to see co-operation and copartnership progress together on the land.

CHAPTER VI

FROM CAPITALISM TO CO-OPERATION IN LEICESTER

After the Rugby conference, I went to Kettering and Leicester, which are strongholds of co-operation and labor copartnership. In Leicester, which has been known for generations as a center of radical fearlessness, are the mammoth shoe works of the Co-operative Wholesale Society of England, called the Wheat Sheaf Works; and not far away are the Equity shoe works. The Wheat Sheaf is run on capitalistic, and the Equity on labor copartnership lines. The latter was started by workingmen who withdrew from the Wheat Sheaf after a strike there. The opportunity to compare these illustrations of two opposite methods was of itself enough to make a visit worth my while; but there are also at Leicester a copartnership printing shop, a hosiery manufactory, an engineering works, and another shoe works.

Leicester is a large city, and co-operation is only one of many things there; but Kettering is merely a town, and its co-operative store and works are its biggest things. The people there live and breathe in an atmosphere of co-operation. After I had gone through the three kingdoms, Kettering stood out in view above all other places as the one which, on the whole, gives the best object-lesson of the spirit and practice of copartnership, though at other places there are copartnership works which are older and as successful. At least half the town is co-operative; and co-operation is in the hands of a group of young and devoted men, believers in labor copartnership not only as the best business, but as the necessary morality of industry.

The English Co-operative Wholesale Society is a society of societies, which was formed to buy in the wholesale markets or manufacture some of the supplies of a large number of the co-operative societies throughout Great Britain. It was begun in 1864. Individual shareholders are excluded. Only co-operative societies can be members. Such an organization had been agitated for many years at the co-operative conferences of Lancashire before the decision, almost unanimous, was made to go ahead. The different societies had found themselves seriously embarrassed in attempting to make their purchases singly. As was explained to me by the very able business man who is now at the head of the Wheat Sheaf Shoe Works, the co-operators went into wholesaling because the wholesalers from whom they had been buying, and who also were supplying retailers competing with the societies, would sometimes refuse to sell, or would charge, whenever they could, a discriminating price. The retail societies saw that their only escape was to confederate and buy such large quantities that no one could or would ignore them. Even where there was no discrimination by the manufacturers or wholesalers, it was good commercial policy for the co-operative societies to unite and buy the commodities they required with ready money in quantities sufficiently large to command the best markets.

The Wholesale was a success in merchandising from the very beginning. It then went into production because, in the flush times following 1870, the manufacturers of Great Britain, overwhelmed with business, supplied their regular customers in preference to the co-operative societies, though glad to sell to them in slack times. To protect its members—the co-operative societies—the Wholesale began to manufacture its own goods, and the same policy has been adopted by many of the other co-operative societies throughout Great Britain.

I was shown all over the Wheat Sheaf Works by the manager, Mr. John Butcher, now a fine type of the modern business man, and forty years ago one of the most active of the co-operative propagandists. The works were built only after he had visited America to inspect all the newest inventions in shoemaking machinery, and required the purchase of six acres of land. One room contains about an acre and a half of floor space. The arrangements are admirable for light, ventilation, sanitation, and the despatch of work. The architectural effect is very good; and the whole fully merits the enthusiastic description of it given by Mr. Benjamin Jones in his work, "Co-operative Production," as the "finest boot factory in the kingdom." In the *Annual* issued by the English Wholesale for 1898, it is stated that the first year's production in 1873 was 90,000 pairs, valued at $118,600. The production of 1896 was 1,237,701 pairs, valued at $1,415,160. The balance-sheet for the quarter ending June, 1897, shows that the production was then going on at the rate of 1,600,000 pairs a year, at a valuation of about $1,700,000. For this quarter, the increase in the production was 55,100 pairs over the corresponding quarter of 1896, and the increase in value was $58,245—an increase of fifteen per cent. The profit for the six months was $33,-

225. There have been two strikes in this factory, as has been told by Mr. Jones in his book. The hours at the Wheat Sheaf are fifty-two and one-half a week; the usual hours in Leicester are fifty-four. All the men are trades-unionists.

The manager of the Wheat Sheaf was outspoken in his condemnation of the labor copartnership idea. It was "the creation," he said, "of an aristocracy of labor." In his opinion, the duty of co-operative manufacturers is to get the goods to the consumer at the cheapest possible price, and they have no right to make this price dearer by paying more than the market rate of wages. Mr. Jones, in an address which he delivered before the Co-operative Congress at Ipswich in 1889, took the same view. "The worker and the capitalist ought to be paid the market rates of wages and of interest. In remedying the neglect resulting from the fact that hitherto the workers have received too little and the capitalists too much in the world of industry," Mr. Jones cannot see, he says, "that it makes any difference whether it is done by giving the worker more wages and the capitalist less interest, or whether it is done by the workers having a share of profits after capital has received its minimum rate of interest." But by this "share of profit" Mr. Jones does not mean direct participation by the workers in the results of their own production. He favors only such profit-sharing as the workingman may secure by membership in the co-operative store owning the workshop. He thereby becomes a sharer in its dividends, which include the profits of the shop he works in.

The debate between other co-operators and the copartnership people is an old one. It broke forth at the very establishment of the Wheat Sheaf Works in 1873. It was proposed by Mr. E. O. Greening that the working people should be made partners and sharers in the profits which they created. "They must also have the right to invest their savings in the concern, and have votes in its management." A very good statement of the argument on the other side was made by Mr. Nuttall, in the debate which followed. "The better policy," he said, "was to let every worker be a member of the (co-operative) store, and let the store make (manufacture) what it sold. He was then his own producer and would receive everything back in the form of dividends, and would be better off in the long run than if engaged as Mr. Greening proposed." The English Co-operative Wholesale gave a bonus to labor at first in its workshops, but abolished it in 1876.

It was after the strike in the Wheat Sheaf Works in 1886, which was bitterly lamented as a scandal to the cause by co-operators, that a number of the men went off and engaged in the organization of the labor copartnership factory already spoken of — the Equity. One hundred of these men were present at the first conference, and they were able to subscribe at once $750 in small sums. They did not have, naturally, much of the good will of the Wheat Sheaf. One of them told me that the manager of the Wheat Sheaf requested several big leather houses not to sell to them, adding that if they did the Wholesale would close its account. But the manager denies this.

This new factory was called the Leicester Co-operative Boot and Shoe Manufacturing Society, known as the "Equity," for short. It started in an insignificant little shop in one of the back streets, but now occupies a very fine building, though not nearly so large as that of the Wheat Sheaf. I was shown all over the works by Mr. John Potter, the president, who pointed out, with special pride, the American system of ventilation which they had; the light and airy rooms, and the many provisions made for the health and comfort of the workers. At the top of the building is something new in factories—a large hall for educational and social uses. It will seat two hundred and fifty people. There are newspapers, games, a piano of co-operative make, a library, portraits of prominent co-operators, and co-operative curtains at the windows. Lectures, entertainments, tea parties, and monthly meetings are held here. This society owns all its buildings and land. Its relations with its employés—or, rather, their relations with each other—are not like those of the ordinary factory. There has never been a strike. Every male adult belongs to the trades-union of the industry. The society has always paid full trades-union wages, and is now paying about five per cent more, including bonus. For many reasons, the co-operators favor the trades-unions. As Mr. Potter expressed it, "This assists our competitors to keep up the same rate of wages we pay. Ever since we started," he said, "we have never cost the trades-unions a cent, nor have we ever had any trouble with them. We hold aloof, too," he said, "as all co-operative societies do, from federations of employers. Ours is strictly a workmen's movement."

I asked what was done with workmen who were insubordinate or inefficient? The manager, he explained, has power to suspend, until the sub-committee can act. If that ratifies his action, the suspended member still has an appeal to the Conciliation Board, composed of delegates from the co-operative societies and the trades-unions. There have not been ten suspensions in ten years in this establishment. In fact, only two or three men have been discharged, and in every case where there has been an appeal, the Conciliation Board has sustained the action of the manager and the subcommittee. Only half a dozen men altogether have left the society since it was formed ten years ago, and some of these have gone to become managers of other societies, such as the Anchor Boot and Shoe Works, in Leicester; also a copartnership shop, and the co-operative societies at Hinkley and Barwell. The advocates of copartnership are fond of calling attention to this fact: that the workingmen went from the Wholesale's capitalistic works to the copartnership shoe works, and that only one of them has gone back. The wages paid are from $4.50 to $6.25 a week for women, and from $7.00 to $11.25 a week for men. The hours are fifty a week. There is a schedule of fines, but it has never been used.

Every workman is made a shareholder in the factory by the retention of his bonus and its accumulation until it pays for the minimum amount of stock necessary. Every workman

WHERE THE "EQUITY" SHOE WORKS STARTED IN 1887
"An insignificant little shop in one of the back streets"

"EQUITY" SHOE WORKS IN 1894
"Every workman a shareholder".

must, therefore, be a shareholder, and every workman or workwoman is eligible to every position in the management. One of the workingwomen was once elected a member of the board, and fulfilled the duties of the position to the satisfaction of the members. Like many other co-operative societies, the Equity shoe works have had more capital offered to them than they could keep profitably employed; and just before I was there, the directors had temporarily closed the capital account. Middle-class people were attracted by the merits of the stock as an investment, and were beginning to subscribe sums of $500 and $1,000, either for shares or for loan capital. In his speech at the opening of the new factory in 1894, Mr. Holyoake said: "This society has $55,000 of capital. Why, fifty years ago, had a few unknown workmen asked to be entrusted with $55,000 of capital, they would have been told they would not get it in eleven thousand years. Nor would they now, were it not that workingmen co-operators all over the country have trust in the honor and ability of men of their own order." When this society was organized, there was no shareholder besides the actual workers in the trade except the local branch of the trades-union. This took $500 of the stock. The co-operative store in Leicester took $125 worth of stock a year later. In the first two years there were only three investments in the stock by individuals, amounting altogether to $287. During this period, when there was practically no outside capital, the society made profits of over thirty per cent of its share and loan capital, and paid its stockholders dividends of eight and nine per cent. There is now a considerable amount of outside capital invested, but this did not come in until the workers had proved their capacity to make the enterprise successful.

The statistics of the Equity works are interesting. In 1887 they had 220 members and $2,100 capital. Their trade the first year amounted to $14,000, and their profit to $1,150. In ten years, at the end of 1896, the number of members was 1,064; the capital, $98,125; the reserve, $4,405; their trade, $235,760. The total profits made by the society in these ten years were $58,745. Of this, the workers had received as their share, $20,310; a dividend had been paid customers of $10,475; capital had received as its bonus $6,305; $2,550 had been spent for education; $5,025 for provident fund for old age and sickness, and $672 for charitable and propagandist agencies. The arrangement for the division of profits is forty per cent to customers, thirty-five per cent to workers, seven and one-half per cent to share capital, five per cent to the educational fund, and twelve and one-half per cent to the provident fund. The provident fund promises old-age pensions for the workers without taxing their current income. The number of workingmen employed is about four hundred and fifty. For the last half of 1897 the trade is reported at $105,905, and the net profit at $4,607; $325 was added to the reserve; five per cent was paid upon capital; labor was allotted sixpence in the pound on its wages, customers two and one-half per cent in the pound on their purchases, and share capital a penny in the pound in addition to its usual five per cent. At the beginning of 1898 the members found that they had to enlarge their factory, and will do so at a cost of $5,000, increasing their capacity seventy-five per cent. The work will be done for them by the Leicester co-operative builders.

As has been the practice of a large proportion of the co-operative societies, special attention has been given by the members of the Equity to providing money for members to build homes. Several building societies have been formed within the membership of the shoemaking factory. About sixty houses have been thus built. I was taken to see several blocks of these houses, and they were tastefully built and conveniently arranged. They have named one street on which some of these houses stand, in honor of the works, "The Equity Road." There are seventeen houses on one side and eighteen on the other, and their average cost is $2,000; but the two houses on the corner cost $3,700 apiece, and three houses adjoining these cost $3,000 each. A $2,000 house contains four bedrooms, bathroom, parlor, dining-room, kitchen, and scullery. There are marble mantels, attractive woodwork, and gas fixtures. These houses, built by the society, have, almost all of them, been sold to employés or "members" of the Equity works. A buyer pays $350 or $400 down, and then pays from $2.50 a week upward until the balance is liquidated.

These shoe workers, as was explained to me by one who had taken the lead in the building enterprise, became dissatisfied with the kind of accommodations they obtained by renting. They had done so well by becoming their own employés that they asked themselves, he said, the question: "Why should not we be our own landlords and have houses of our own to suit ourselves?" To my suggestion that it was a risky business for men on weekly wages to undertake the responsibility of purchasing houses so good as these, he replied: "The society has,

LEICESTER'S WORKINGMAN'S HOUSE
"Why should we not be our own landlords?"

of course, legally, the right to discharge one of the workers at any time it thinks best; but, morally, it has no right to do so unless he misconducts himself." In the co-operative world, the "right to work" has become a reality. Standing under this sort of dispensation, the workingman who behaves himself knows that he can take greater risks than the ordinary employé. In constructing the houses which I saw, the building society had employed a contractor; but the new houses are being put up by a co-operative builders' association. The houses are not on leaseholds, but on land which the society owns in fee.

A few minutes' walk takes us from the "Equity" to the Leicester Hosiery Manufacturing Society. This concern is now finishing a large modern building, which will increase its manufacturing capacity to $450,000 a year. It began twenty-one years ago—they are to have a grand "coming of age" celebration this year—on a capital of $150, in the front room of a cottage, for which a rent of twenty-five cents a week was paid. More than half of it belonged to the employés, and nearly all of them were hand-frame workers. Its present manager was a frame-worker, and his associates were men who, like him, had been driven to seek some co-operative method of employing themselves, since the labor troubles of the time and the slackness of business had left them out of work. In the first two years, they had to borrow almost continually, as they were short of capital and were glad to get a month's credit for yarns. They had no money to spend on traveling men, and got their trade by sending samples to co-operative societies. In this way they received many small orders, and in some cases societies that could not give them orders would send back the samples, carriage paid, with some word of encouragement. At the end of the second year, their capital was $2,385. They now have a share, loan, and reserve capital of $195,000. Their sales are running at the rate of about $250,000 a year. Their manufacturing capacity is now over $300,000 a year. From the beginning the concern has doubled its trade nearly every four years. When the building is finished which is now being constructed, and is filled with machinery, it will be able to manufacture nearly half as much again.

In one of its leaflets is printed an interview in the *Workmen's Times* with Mr. James Holmes, secretary of the Amalgamated Hosiery Operatives' Union. "They can show you," said Mr. Holmes, "a list of names at this place as long as my arm, of people who are waiting

to be taken on; but the mischief is nobody ever leaves, and it is only the vacancies caused by death and the new situations created by the extension of the business that afford openings. One thing that makes the place such a valuable one at which to work, is the fact that employment is very regular. They cater for a special line of customers, the distributive co-operative societies all over England, Wales, and Scotland; and the varieties of goods required by these societies are so great that there is nearly always work on hand, and if there is no order on the books, it is nearly always safe to work to stock. You may guess, therefore, that there is always a desire to get employment here, and that very seldom is there a case where a man or woman once employed leaves the place." I found two hundred and fifty employés in the building, and there are fifty on the outside.

The men and women are all shareholders. It was formerly optional with the employés whether they would become shareholders or not; but under new rules which have been adopted, every employé must become a shareholder and will receive none of the profits in cash until $50 has been accumulated for the purchase of his stock. Of the nine directors, two must belong to the factory; and, accordingly, at the meeting of stockholders in February, 1898, two of the employés were elected directors. "Workers come here," the manager said to me, "and do not go away. We do not look upon the business as merely commercial. In buying our supplies we get them as cheaply as we can, provided they are produced under proper conditions. One house that sold to us boasted to me that none of its people were in the trades-union, and that was the reason he could deal with us more cheaply than others could. We closed his account *instanter*. If we in the co-operative industries stand for anything, we stand for a better order."

The method by which co-operative societies get more capital when they want it was illustrated recently in the case of this hosiery concern. Finding that their growth made it necessary for them to use more money, they issued an appeal by circulars sent to the societies that were buying their goods. This was in April, 1897, and by July they had received subscriptions of $41,275, for shares, and $14,395 for loan.

When I visited these hosiery workers, they were in a flutter of apprehension that they were about to be swallowed up, or, "competed out" of existence, by the English Co-operative Wholesale, which has something of the reputation of an octopus. In a circular the hosiery society call attention to a motion which had been introduced at the quarterly meeting of the English Co-operative Wholesale Society, directing its management to commence the manufacture of hosiery. The circular went on to recite the facts we have given with regard to their growth, and the fact that their new extensions gave them an increased capacity, ample for the present needs of the trade; it pointed out that the Wholesale society, which takes one-quarter of their entire product, and 294 other distributing societies, were stockholders in their enterprise, and were, therefore, deeply interested in the proposed invasion of their field by the Wholesale, and submitted the matter with this very gentle suggestion: "We are of opinion that if this motion is carried, the Wholesale would have to put down a plant to make the same class of goods we are now satisfactorily supplying to it. Hence, there would be two plants of expensive machinery competing against each other, the money for doing this coming from the same source. We commend these facts to your careful consideration, and hope you will take such action as you consider best for the movement as a whole." The result was a vote against the proposal. In a number of cases the English Wholesale society has entered into competition with other co-operative societies which had successfully established a productive business.

The Leicester Co-operative Engineering Society was started by two working engineers, who found themselves, in 1893, out of employment. They resolved to make the experiment of employing themselves if they could get the support of the other co-operative productive societies in the town. The answers were encouraging enough to induce them to go ahead. They formed a society and began their subscriptions of a shilling a week. They were given a room to meet in by the Leicester co-operative store. They began work in 1894, and have been somewhat embarrassed because they started—as many of these societies do—with insufficient capital. The constant exhortation of the organizers of co-operative production to new beginners is: Don't begin without being sure you have enough capital. The other societies in Leicester gave them what work they could, and they are, at last, standing upon a basis of solid prosperity. Their working staff now numbers eleven, all of them shareholders but one; and three, including the manager, are committeemen. They were able to show a profit last December for the first time, and succeeded in wiping off previous losses. When I was there, these men had, each of them, contributed $6 to the support of the Amalgamated Society of Engineers in their lockout struggle with their employers. This was their extra levy for one quarter besides their regular dues as members of the A. S. E. And they would gladly, they told me, pay twice as much.

The Co-operative Printing Society of Leicester is an important institution in the copartnership world, as it prints the organ of the movement, *Labor Copartnership*, and its documents. This society has been in existence four years, and has made a profit all through with the exception of one half-year. It owns its own machinery, and has twenty to thirty members. The managing committee is chosen partly from the workmen, and its members get, besides their pay and their share of the profits, an allowance of ten per cent as compensation—an arrangement which is usual throughout the co-operative world. A member of a managing committee in a co-operative concern like this gets, first, his wages if an employé; second, his share of forty per cent of the profits, according to his wages; third, an extra payment as member of the committee, which, if there are, as here, nine members, would be one-ninth of ten per cent; fourth, his dividend on whatever share capital he may have in; fifth, his interest on whatever loan capital he may have deposited; sixth, his share of the benefits of the provident and educational fund; and, last but not least, his general benefit as a co-operator.

CHAPTER VII

KETTERING

There are co-operative works in England older than those at Kettering and equally prosperous, as at Hebden Bridge; but there is no other place in the country where, on the whole, copartnership is so well illustrated. The radicalism of the early part of the century prepared the ground on which co-operation has built, and the pioneers here in the last generation were men who, clearly, got their inspiration from the seditions in religion and politics which were so rife in England in their day. Kettering is especially interesting to the co-operative traveler because here co-operation has followed a normal course from distribution to production, and has, moreover, maintained its faithfulness to the highest views of the founders. First, a distributive store; second, workshops, in which the workers are part owners and directors; third, homes for co-operators; and, fourth, a farm. This has been the line of development. The Kettering store has been a reservoir, gathering profits for production and distributing them as capital to open new avenues for the employment of labor. Its members have seen that every productive enterprise that was planted by its side has been a new buttress for the main building. In going to the farm of which we have spoken above, the Kettering co-operators are but completing the circle of co-operation, and tracing the lines on which the co-operative commonwealth must grow. Its broadest side must rest on the land.

In Kettering may be seen a co-operative store which has managed to absorb all the capital which it has produced and more. It is borrowing to extend its productive enterprises from the Wholesale, which acts as a banker, as well as a wholesaler, for many of the co-operative societies of Great Britain. Meanwhile, other co-operative societies in England, not forward enough to use their commercial profits in the establishment of manufacturing, are forcing their surplus capital on the Wholesale to be invested in consols or something not so good. Stores as progressive as this at Kettering are not yet the rule. It is quite a usual thing for a co-operative store to be found opposing the development of productive works, or to be at least neutral. The capitalistic instinct seems to come with the possession of capital, and prefers to send its money to be invested in government bonds, rather than take the risk and undergo the agitations of pioneer work in co-operative production. This work always requires an appeal to moral considerations and to sentiment, and the average business instinct, even among the co-operators, has not yet learned that, of all the forces that create and determine values, the most important are the moral and sentimental.

The hard times of 1865, and the example of other co-operative successes, were the immediate causes of the start in Kettering. Wages were then low, and the majority of the people found it hard to make both ends meet. There had been a co-operative store attempted some years before, but it had come to grief. But co-operators in England have never shown the weakness or the lack of intelligence to be discouraged by first failures. I quoted to Mr. E. O. Greening, author of the "Co-operative Traveler Abroad," and one of the leaders of labor copartnership, the conclusion of an American friend who had tried to organize a co-operative store and failed, that "co-operation was no solution in America." "We got through with that sort of talk here thirty years ago," he said. "We owe our success to the fact that after failing and failing time after time, we would always begin again." Mr. George Jacob Holyoake, too, said that no conclusion could be drawn from American experience except to keep on trying. "The argument that co-operation cannot succeed in America is as baseless as the cry I remember twenty years ago—that America and New Zealand were overpopulated. But they now support many millions more, comfortably, with room for uncounted millions still." To the remark that some zealous workers in American co-operation felt discouraged and unappreciated, Mr. Holyoake replied, "The co-operator must never care whether he is appreciated or not. He must fix his eyes on the results he wishes to achieve, the wrongs he wishes to end, and keep right on, whether he fails or succeeds, whether any one listens or not."

In 1866 the co-operative store was tried again in Kettering. At the beginning the membership was sixty; the capital, $455; and the store was one room in a dwelling-house. One of the workingmen was appointed storekeeper. A signboard, "Kettering Industrial Co-operative Society, Limited," was put over the door, and the members began to buy their groceries from their own shop. They entered with pride and confidence upon another experiment in the English policy of social emancipation by the prosaic device of buying and selling their own groceries among themselves, and saving the middleman's profits on their scanty rations.

The first quarter's trade amounted to $2,030. This was not bad considering that only two members were well enough off to pay up their contribution of $10 of capital, and all the rest were paying it at the rate of only twelve to twenty-five cents a week. Still, the committee were able to divide a dividend of nearly four per cent to the members on their purchases. There was a great jump in the profits of the second and third quarters' trade. The dividends rose to one shilling sixpence and one shilling eightpence on the pound.

The members were so delighted that the secretary was instructed, by resolution, to send a report to the local newspaper, *The Midland Free Press*, but at the end of the year the discovery was made that in these quarters, through some mistake of the workingman who had the figuring of the accounts, what had been divided among the members had been the entire bank balance instead of accrued profit. This made it necessary to cut the dividend down sharply, and the result was serious disaffection and a diminution in the trade of the society, although a majority of the members remained loyal.

From these humble beginnings the society has developed a business which is now running at the rate of $400,000 a year. The society has grown from seventy members in 1866 to 3,983 in 1897. This has been the fruit of self-sacrificing labor on the part of the founders. They had to attend to the business of the store after working all day at their trades; and one of the early pioneers remembered with especial affection is Mr. Edmund Ballard, who used to devote every evening of the week to the work of the society, sometimes until midnight three or four nights of the week, returning home too tired to sleep. He was "anxious," as he often expressed it, "to do what little he could toward leaving the world better than he found it." These men were sneered at by their non-co-operative neighbors as "the shoe-making grocers." Their only response was to buckle down all the more earnestly, to prove that they could be successful grocers even if they were shoemakers. The society has just completed the construction of a block of buildings for its stores on the main street of Kettering, which, when they are finished, will be the finest retail premises in the town. With the completion of these buildings, the society finds itself the owner of over $225,000 of freehold property, without a mortgage on a single building.

All through the co-operative territory of Great Britain one finds that the co-operators are laborers in the struggle between labor and capital, who have the unique experience of being embarrassed by too much capital. At the meetings of the Kettering society, one of the principal subjects debated is the difficulty which the society has with its surplus capital. In one of their reports the managing committee said: "Your committee wish to call your special attention to our share capital, which has increased $30,000 during the last year, and now stands at over $200,000. While we are pleased to see the confidence placed in the stability of the society by its members, your committee experience difficulty in investing the capital at anything like the interest paid to members." It had bought a large amount of property. It owned twenty stores and other business premises, and a hundred houses, which it had built to rent to its own members; and on all this property no one held a cent of incumbrance. It had advanced money to start various productive enterprises, and was just entering upon the purchase of a farm, and yet had a surplus of capital. On this surplus five per cent interest was being paid to members to whom it belonged. The great difficulty experienced in investing the money at anything like that rate made it necessary either to return the capital to the members or to reduce the rate of interest from five per cent to four, and the society unanimously decided in favor of the latter policy. It was also decided to lend out money to members on the security of land and buildings at four and one-half per cent interest, to be repaid by installments. By January, 1898, the society had advanced over $40,000 to its members on freehold property.

The question of farming was taken up in 1895. The managing committee of the store was instructed at a meeting of the members to make investigations as to the amount of farm produce required by the store. It was shown that in the preceding year they had sold 119,000 pounds of butter and 430,000 eggs; had slaughtered 236 head of cattle, 690 sheep, and 520 pigs. When a vote was taken as to whether the society should go into the farm, there was a majority of two in its favor. It was felt that this majority was too small to make the society strong enough in so difficult a work as co-operative farming had proved to be, and the project was, consequently, deferred, and had not been taken up again until the time of my visit. A very much larger majority being by this time convinced that the venture was a wise one, and was a necessary part of the full scheme of co-operation, it was resolved to go ahead.

The store has a penny bank for the children, in which there are over 4,000 depositors, and in which during 1897 over 50,000 deposits were made. It has also a coal department; three bakeries; a butchery, which kills from 1,500 to 2,000 head of stock a year; and since 1883 it has had an educational department, which is considered an indispensable part of the work of every co-operative store. One per cent of the profits is given to it.

An instance of the methods of co-operative financiering was the purchase of gas stock, made as an investment by the co-operative store. They paid $4,800 for it. In accordance with their practice of "depreciation," they wrote something off its value every year until it stood on their books at only $3,000. Then they sold it for $5,500. Their reasons for selling it were interesting. They found that the workingmen who went to the stockholders' meetings to represent the society were not treated quite in the same way as the other stockholders; and as they did not propose to submit to anything of that kind, they sold their stock and added the proceeds to the capital, which they hold to lend to their members to buy land and build houses, or to advance to new societies wishing to go into co-operative production.

The main store opens at least one branch every year, and has opened eight in the last five years. It has fifteen of them altogether in Kettering, and owns the land and buildings. The store is the largest subscriber to the funds of the hospital, and will soon have a co-operative bed for the use of its members.

The co-operative societies of Kettering have a membership of 4,000 out of a population of 25,000. This means that fifty per cent of the people of Kettering are co-operative. Rugby is still stronger; and there are places like Desborough which are practically all co-operators. The store has published a souvenir, handsomely illustrated, with the title of "Co-operation in Kettering from 1866 to 1896," and bearing on the cover that favorite motto of the movement, "Each for all and all for each," in which these and other facts of its history and growth are given. In this book the spirit of the co-operative movement is so well presented in one or two passages that they may be worth reproducing here:

"We are pleased to know," the writer of the

sketch says, "that our members are beginning to realize that the future prosperity of our society and co-operation in general lies in the extension of co-operative education and of co-operative principles, and that there is a grander ideal in co-operation than mere shop-keeping."

* * * *

"Co-operation has done great things during the past fifty years, but we believe if co-operators will study the broad principles on which it was founded—'Each for all and all for each'—the next fifty years will do still more to spread through the country greater blessings than it has done in the past, and show to those who now look upon co-operators as their enemies that our aims are, after all, to raise the workers in the social scale, and to help them to realize that life is worth living and, ultimately, to form one grand brotherhood."

* * * *

"A great work has been accomplished—how great the figures alone do not include; but there is still room for a great deal more to be done. It has been said 'Co-operators know no finality.' We look forward to the future of the movement with confidence. It is recognized to-day as one, if not the greatest, of the democratic institutions in the country, and is capable of achieving still greater success in the future than it has done in the past, and we trust that the members of our local societies will be prepared to do their part in this progressive work." Paragraphs of this kind seem altogether in place in annual reports of the business of corporations engaged in co-operation; and when they are accompanied by statistics of such steady and progressive prosperity, they suggest that the political economy of selfishness has a thing or two to learn.

The first productive enterprise entered upon after the Kettering store had proved to be a success was the Kettering Boot and Shoe Manufacturing Society. This owed its start directly to a few ardent co-operators inspired by the success of the store, and of the Equity boot and shoe works, of Leicester. Ketteriug has always been a boot and shoe making town, and it was decided to venture in this industry their first experiment in production, in which the worker should not only have part in the management, but also part of the profits he created. The start was made at the end of 1888, with one hundred and forty members subscribing a capital of $2,000, and it was resolved to manufacture exclusively, if possible, for the co-operative market. The result of the first quarter's operations was awaited with interest, not only

by the members of the society, but by the cooperators of Kettering generally, as it was felt that more was involved in this experiment than the mere question of a profitable investment for surplus capital. During this first quarter the new factory opened up accounts with seventy societies, did a business of $4,340, paid interest on its share capital at the rate of five per cent, and made a profit of $23.50. Instead of being divided, this was made the nucleus of a reserve in accordance with the almost universal cooperative practice of accumulating a surplus in good years to meet the losses of bad years. The factory which had been taken became at once entirely too small for the requirements of the business, a piece of land was bought and a four-story building put up at an expense of $7,760. In a year larger premises again became necessary, and the society bought a plat of ground with a factory on it, at a cost of $17,500.

"It is all ours," said Mr. F. Ballard, as he showed me over the fine building. "When we first started," he said, "we were abused as fanatics and, in a quiet way, were bulldozed. Other manufacturers at first refused to sell us supplies, though now they are only too glad to do so. Out of our $50,000 share and loan capital, $17,500 is owned by our own employés, men and women." I asked if there had ever been a woman on his board of directors. He said, No, and added: "The women never nominate a woman." The society had been, he said, "a great success from the very start. One cause of this is that it makes, practically, no losses. It sells wholly to co-operative societies on fourteen days' time; and the losses in the nine years of its existence from bad debts have not been $200." It now has six hundred and eighty members.

A doubt instinctively rises in the mind of the ordinary business man as to the feasibility of running a business by universal suffrage; but in this concern there has never been an opposition ticket for directors, and the present manager has held his position from the start. Of the six hundred and eighty members, four hundred are resident members; and of these four hundred, about two hundred attend the annual meetings. There is seldom a close division in the meetings. The fact that there was once on the question of a purchase of improved machinery a majority of only six out of a vote of four hundred was a circumstance recalled as unusual.

The society puts in all the improved machinery that appears; but its business has also increased so steadily that it has never had to discharge an employé. Like the store, it is also accumulating a surplus, and just before I was there had made a loan of $5,000 to a co-operative building and contracting society, recently begun in Kettering on copartnership lines. All through the co-operators of Kettering, it seems to be accepted as the only sound political economy to invest as much capital as is prudent in productive enterprises, both as new illustrations of the co-operative principle, and to give the co-operative stores further security by establishing independent sources of supply for the manufactured goods they deal in. "We object," said the manager of this boot and shoe factory, "to the policy of leaving our money in the banks. Our competitors borrow from the banks, and hence, co-operative societies which deposit their money there are simply supplying the sinews of war to the other side."

Work in the factory has been continuous since it began. There has never been a shutdown. The society has worked short time only four weeks in its ten years. Ordinary manufacturers shut down two months every year. The working hours are forty-eight a week; other manufactories work fifty-four. Trades-union wages are paid always, and there has never been a cut. When the trades-union rate advances, it is a point of pride with the co-operative manufacturers to be the first to accept it. The society has never had a strike. There was a great strike among the boot and shoe works in 1896 all over the country; but the Kettering co-operators kept right on and not only were able to retain in employment all their own people, but gave work to a large number of co-operators from the outside.

"Ninety per cent of our members," said Mr. Ballard, "are trades-unionists; and even those who are not are glad to pay levies." This factory adopted the eight-hours system voluntarily in 1894—among other reasons, its manager said, "to help the unemployed." The wages of the men are from $6.75 to $8.75 a week; those of the women, $3.00 to $3.75. There is only one piece-work department, and this the manager intends as soon as possible to change to day-work. Two hundred employés are shareholders in the factory. They get, with their bonus, more than trades-union wages, have to work only eight hours a day instead of nine, and have their vote in the management. More and more of them own their own homes. The store has been building houses to rent, but is now beginning to sell land to the co-operators and lend them money to build with.

"Our stockholders," the manager said, "are

almost, without exception, working people. The co-operative movement has the cream of the working class; but our workers," he admitted regretfully, "are a little apathetic in their interest in co-operation and social reform on account of the prosperity which co-operation and social reform have brought them. When a co-operator gets up in the world, he is apt to forget about co-operation. All the Kettering co-operative societies—and there are five of them—have united in educational work to strengthen the propaganda to keep those in the movement alive and to bring in new converts. We want to get the working people to feel as if they had some interest in something besides themselves, and to teach them that business can be organized to make men as well as money." The business of the society has increased, until at the end of 1897 it had a capital of $52,500; a trade for the year of $164,620, an increase of $14,000 in spite of the unsettled state of business; it made a profit in the last half of 1897 of $4,860; its reserve fund is $3,525. It continues to pay capital seven and one-half per cent in interest and profit, and labor also seven and one-half per cent in addition to its trades-union wages and the eight-hours day, after writing off for depreciation two and one-half per cent for land and building, and ten per cent on fixed stock and machinery. The shoe factory has, from the beginning, been an annual subscriber to the Northampton General Infirmary and the Kettering Nursing Association; has given $300 to the general hospital of the town, and $125 toward the Kettering free library, besides maintaining its own educational work.

I was surprised to hear its manager complain of competition from other co-operative manufacturers. At a co-operative conference at Leamington, one of the district store managers said that he had had five "drummers" from co-operative boot and shoe manufacturers call upon him in one morning. This shows the competition there is inside of the co-operative movement. It is now under discussion to federate the numerous co-operative boot and shoe factories, and to do all their selling through an agent.

I take at random from one of the balance-sheets of the Boot and Shoe Society this item in the Educational Committee's report, as something out of the ordinary run of business literature, but seeming altogether in place in co-operative bookkeeping: "Co-operative Sunday was observed on May 2d, with its usual success, our contingent marching from the factory to the Victoria Hall, headed by the Kettering Rifle Band. Special hymns were sung, and a stirring address was given by Mr. G. R. Thorne, of Wolverhampton, on 'The Fatherhood of God Involved in the Brotherhood of Man,' and on co-operation as a realization of the latter; a collection being made on behalf of the Kettering Nursing Association.

"A children's festival was held on Saturday, July 17th, when upward of fifteen hundred children of co-operative members assembled at the Rockingham Road Board schools, thence marching in procession through the town to the North Park."

When the Kettering Clothing Co-operative Society opened its handsome factory, the local paper of Kettering, *The Leader and Observer*, gave half a page in its news columns and an editorial to the event. This new copartnership enterprise had its immediate suggestion in a boycott of the co-operative store, attempted in 1893, by a firm of manufacturers in Kettering. The store had added a clothing branch to its other departments, and sought to purchase its supplies from these local manufacturers. But they would not sell because the co-operative store was the rival of one of their principal customers in Kettering. Some of the employés of this firm were members of the co-operative store; and upon this refusal they said to their fellow-members that if assisted with capital they would organize a co-operative manufactory to supply the clothing needed. A meeting was held, which was addressed by Mr. Henry Vivian, secretary of the Labor Association, and it was resolved to go ahead with the project. Thirty of the audience put down their names for stock and undertook to solicit subscriptions from others.

Seven of the workingmen who thus took shares were employés of the boycotters. This firm at once served notice on them that they must either give up all connection with the co-operative society or lose their places. The workingmen stood up manfully, and left their employment rather than leave the co-operative society, though in the home of the leader among them was a young wife with a new-born baby. As one of the speakers said at the opening of the new factory, as chronicled in the newspaper, "If it had not been for the action of that private firm, the town would not have had this new society." The start was made with twelve workers, seven of them the men who had been discharged from the local firm. They were drilled in co-operative ideas and methods for twelve months by Mr. W. Ballard, of the store, before beginning.

There were fifty-eight other shareholders, who were mostly workingmen in the private works, waiting until the business grew large enough for them to find employment in it. One of them said to me: "In the private works we saw that we were fairly well paid, but it was only while we were young. As soon as one got old, he was laid to one side. This kept us thinking how to better ourselves, and we were all ready to come into co-operative production when this opportunity presented itself." In their first shop they began work under a zinc roof, through which the wind and rain drove in; it was very cold; the need for economy compelled them to be sparing of fuel; they had to work early and late, and sometimes could not help wishing they were back, as one of them said, "in Egypt."

The manager, referring to the same apathy among the members, of which the manager of the boot and shoe works had spoken, said: "We would be better off now, if, instead of there being among us those seven who had roughed it, we had seventy. Our new members know too little of adversity."

In the first half year of 1894, business was done to the amount of $11,375, with a profit of $1,170. Like the boot and shoe society, the clothing society found at once that the shop it had taken was too small; and it had to move four times into larger premises in the first two years of its existence. The building in which I found it, the opening of which was noticed with such display in the local press, was a fine four-story brick structure on one of the principal corners, with large windows, the best of ventilation, labor-saving and sanitary appliances, heated with hot water and lighted with electricity. It forms a striking contrast with the first factory in which they established themselves; which was a very small two-story building, not much better than a shed, on one of the obscure streets of the city. For their new quarters, the engineering work was done by the Leicester Co-operative Engineering Society, already referred to, and the building by the Kettering Co-operative Builders, on land which was bought from the co-operative store, which also advanced funds for the building. Mr. William Ballard, the manager of the store, says: "'Help one another,' ought to be a part of our creed as well as self-help."

In their souvenir of the dedication of this building, the society says: "The factory, besides being modern in construction, is itself a sample of co-operative production of no mean type" The building is one of the most conspicuous in Kettering. Its height is sixty-four feet to the ridge of the roof. It has a

KETTERING CLOTHING FACTORY IN 1893

OPENING OF THE NEW KETTERING CLOTHING FACTORY IN 1893
"The Helping Hand"

decorative entrance, and over the front doorway is carved the emblem of the society—"A Helping Hand" — the device being a man extending his hand to help another upward. Upon the top stone is engraved the name of the society, the date of its formation, 1893, of the erection of this building, 1895, and the name, "Excelsior Works." There are two separate entrances for work people, so that in case of fire there would be ample means of exit. The staircases are fireproof.

The society began with a dozen workers and a capital of less than $2,500. At the end of 1897, the capital was $44,855—$10,000 of it owned by the workers—and its trade for the year, $123,385. During the last half of that year an increase in their trade of $8,422 is reported over the corresponding period of the year before. The net profit, after paying interest and depreciation, amounts for the half year to $3,380; of which $1,231 goes to the workers as their share, at the rate of one shilling sixpence on the pound. The society has made a success of the eight-hours day, and the profit-share of its workers amounts to seven and one-half per cent on their wages. These working-men have a factory which they accurately describe as one of the finest to be found in the kingdom. Not a man has left the place since they began, and only one or two girls. In going through the works with the manager, Mr. L. Jessop, I was introduced to two young women, who were sitting with a number of others at the working-tables in one of the rooms. They were wage-earners. But they were something more—they were also directors in the board of management of the society. They were as efficient and business-like in the board room, I was told, as any of the other directors. At this spectacle of women wage-earners, who were also shareholders in the concern which employed them, and were elected by their fellow-workers directors in its management, I marveled much.

Kettering co-operation is already beginning to taste some of the perils of prosperity. At a recent auction sale, at the beginning of 1898, some of the shares of the Kettering clothing society sold at sixpence above par. This is a clear foreshadowing of the danger of the entrance of shareholders seeking a profit alone, and by their presence threatening the co-operative policy of the society. To meet this, an amendment of the rules is being discussed to prevent outside capitalists from speculating in the shares. This can be done by giving the managing committee the power to buy the stock on the death of shareholders, and also to refuse transfer at any price above par with compulsory powers to buy shares.

The Co-operative Building and Contracting Society of Kettering resulted from the defeat of a strike of carpenters in May, 1893, for an increase in wages. The failure of the men was due to the fact that their trades-union was not well organized. Discontented with the result of their strike, and not being willing to go back to work under the depressing conditions of private employment, if by any effort they could secure for themselves the wages and independence which they saw the co-operative workingmen enjoying, the carpenters appealed to Mr. F. Ballard, the secretary of the shoe factory, for his advice and assistance. The result was the formation of a society to go into the building business. The first contract was for the erection of the new factory of the co-operative clothing society for $15,000. They were not the lowest bidders, but the building committee of the clothing society unanimously decided to give them the contract. The clothing factory procured the money to put up this building by a loan from the co-operative store at the very low rate of four and one-half per cent, and then, in turn, stretched out its helping hand to start the builders by giving them the contract.

The new venture did $30,000 worth of work during its first year, 1895, and each year since has shown an increased prosperity. How closely their manager does his figuring, appears from the fact he gave me that he had lost a $20,000 job by a difference of only $20 in the bids. A large part, but not all, of the work done by the builders comes to them from the co-operative store and the other co-operative organizations of the town. The society now has thirty employés, who are also all shareholders, and has a capital of $16,835, of which its workmen own $2,000. Most of the other stockholders are from among the members of the other co-operative organizations. It is the custom of the co-operative workingmen, and also of the co-operative society, as they get on in the world, to invest their savings by taking shares in numbers of different co-operative concerns. There are a few, however, of middle-class investors in the building society attracted by the ten per cent interest which they have paid; but as soon as the society is a little more firmly established, measures will undoubtedly be taken, as is done now almost universally throughout the co-operative world, to discourage the influx of such capital by a sharp reduction of the rate of interest.

I went through their shop and yard, well

stocked with tools of their trade and every needed variety of lumber and material. They own this building, worth $7,500, all paid for. Co-operative builders societies are found in Oxford, Cambridge, Plymouth, Exeter, and elsewhere. There is a successful one in London, which, like this in Kettering, goes into open competition in the general market. Next to the shop and yard of the Co-operative Building Society I had pointed out a handsome little brick and stone structure, which was the clubhouse of the town band, whose members are recruited from the musical co-operators. All the plans of this building were drawn by the manager of the building society, Mr. A. Bamford, a workingman who had given himself the special training needed for this work. The society has had three successful years, and paid its employés six per cent bonus the first year and ten per cent the next. During 1897 the trade was $38,563, and after paying interest, depreciation and other fixed charges, there was a divisible profit of $3,502. After paying the workingmen a dividend of one shilling ninepence a pound, the society was able to place $565 in the reserve fund.

A new project came to the front while I was in Kettering—the establishment of a copartnership corset factory. I had the good fortune to be present at the meeting which was held to consider its advisability and to devise ways and means. I was eager to see how the new democracy of industry went about its business. The meeting took place in Co-operative Hall, a neat little brick and stone affair, which the co-operative store bought in 1892, and in which the business and educational meetings of the Kettering co-operators are held. It is a general thing among co-operators to furnish themselves with their own halls. They think the care and the expense which they go to in the provision of their own books and newspapers, and library rooms to read them in, the best of investments, as rendering them entirely independent of local influences which might seek to hamper them. At the beginning of the movement, the hostility of the vested and business interests often manifested itself in denying them meeting places, and the censorship of the parson and squire is still remembered and resented. Co-operators began establishing free libraries long before the Free Libraries Act. At a very early period in the movement, co-operation set before itself the task of becoming mentally independent as being quite as important as that of being independent in its groceries.

On this evening the little hall had an audi-

CO-OPERATIVE HALL, KETTERING

"The hostility of the vested interests often denied them meeting-places"

ence numbering about two hundred, practically all of them co-operators. It was a meeting of working people who had become manufacturers, and who came here to help other workingmen to become manufacturers. Women were present as well as men, and young people were numerous. The proceedings followed the usual order of public meetings. A chairman, Mr. F. Ballard, of the shoe society, was elected, and a secretary, Mr. L. Jessop, of the clothing society. The meeting was opened by an address by the chairman, who spoke in a plain, matter-of-fact way to the particular point for which they had assembled. "We have," he said, "to discuss whether there is room in Kettering for a corset manufactory on co-operative and copartnership lines. There is no such society as yet anywhere in the co-operative movement. We have a great many boot and shoe societies, working almost wholly for the co-operative societies, and almost all doing well. The productive enterprises of the movement should be able to look first to the co-operative societies for their markets; and these are able to absorb a very much larger co-operative product than they are now taking."

Though the speech was strictly a business speech, the co-operative idea made it quite consistent with business principles to introduce moral and sympathetic considerations. The chairman evidently thought it to be a part of the question whether they should go into the manufacture of corsets to remind his hearers that they were in morals as well as in business, and that, in his words, "One of the things that co-operative production has to do is to produce a new moral character in the business world." The character of the response of the audience to every reference to moral wealth as a necessary part and preliminary of other wealth, showed how accurately he represented the ruling feeling. "We are," he said, "in the co-operative movement not principally for the dividends, but to help each other. Our duty as co-operators is not merely to help ourselves, but to help our brothers and sisters in their trade, which is their life." He gave very practical advice to those who were thinking of going into the corset works: "Don't start in a hurry; don't start until you have money enough. We need," he said, "$2,000. That ought not to be long in forthcoming." An address was then made by Mr. Henry Vivian, the lecturer of the Labor Association, which was devoted to an exposition of the duty and advantage of co-operators in employing their capital in productive enterprises on the basis of full recognition of a share of the profits to the worker as his right.

After this, one of the audience moved, "That in the opinion of this meeting of Kettering co-operators, it is desirable that a corset manufactory, on a labor copartnership basis, be started in Kettering." This was put to the democracy and was passed by a unanimous vote, in which woman suffrage had its full part. A committee of five representatives of different societies was then appointed to draw up rules and make arrangements, with power to add to their number. The chairman at this point intervened to give a warning to those who were present. "You understand," he said, "that there has been trouble in the past for the employés of private firms who ventured to take part in enterprises of this character. We do not want any such trouble in this matter." This was in reference to the discharge of the workmen who took part in the organization of the clothing factory. "It would be a very undesirable thing," the chairman said, "to have any person's name mentioned here to-night." Three employés of a private corset factory in Kettering who had saved $125 each were known to be ready to come into the co-operative factory as soon as it was organized, and there were others who would follow them; this admonition from the chairman was to protect them from any indiscreet disclosure of their names and purpose in the hearing of the local reporters or casual spectators. A stock subscription list was opened, and the chairman issued a warm invitation to all in the audience to be on the "honor" roll of the subscribers of the first night. Eight hundred and seventy-five dollars was subscribed before the meeting adjourned.

Since then, enough for the start has been raised. A good factory has been found, and, regardless of the prophecies of great political economists that the productive workshop has "no future," the society took pains to rent only a factory which had with it grounds for an extension. The co-operators at Kettering, in going into corsetmaking, have shown themselves, like the Scottish Co-operative Wholesale Society in taking up shirtmaking, confident in the economic superiority of their methods over those of the sweater. In the corset trade, the bulk of the workers are women and girls, always subject to the competition of married women; and the conditions of labor have, therefore, been very bad. The Kettering co-operators pit factory and the co-operator against the sweater with every confidence not only of ameliorating conditions, but of making a business profit.

Even the most "scientific" person could not view the workings of co-operation in Kettering without feeling some glow of enthusiasm. "A real piece of the Kingdom of God actually arrived," one of the Oxford University men, who has made a special study of industrial questions, said in describing his visit to Kettering.

CHAPTER VIII

LOSS SHARING

Mr. Gladstone once remarked that he could understand about workmen sharing the profits, "but how," he said, "about sharing the losses?" That question answers itself in copartnership. The workingman who becomes a shareholder, with his participation in control and in responsibility and profit, becomes equally a participator in losses when any are made. I found more than one copartnership factory, particularly in the depressed cotton districts, where the workers were uncomplainingly bearing their burden of losses as a matter of course. They comforted themselves with the reflection that they were only losing a part of previous profits that they never would have had but for copartnership; and that their collateral gains in business training, education, and fellowship were theirs forever.

At Burnley this "sharing of losses" was to be seen. The whole cotton-spinning country was in the trough of a wave of depression, and the workingmen in the labor copartnership "Self-Help Cotton-Spinning and Manufacturing Society" were sharing losses, not profits. Since my visit the news has come, in January, 1898, that the society has gone into liquidation; but it is announced as certain that its assets are ample to pay off all the liabilities, and that it is hoped the society may be able to continue.

These copartnership works were the offshoot of the Burnley co-operative store—though not with the help of the store, which was decidedly hostile to the new enterprise. But this store had for many years been especially liberal in its appropriations for educational purposes. In 1892, for instance, it spent over $6,000 in this way, and for the year 1897 the appropriation was the same. Out of these educational funds the society gives $250 a year in prizes to the evening schools for attendance. Any co-operative child who has marks of ninety per cent for attendance gets a prize. The store has thirty-two branches. A very extensive system of reading-rooms—of which there are fourteen—is part of the educational work at Burnley. When the co-operative store members in a neighborhood want a reading-room, they submit a petition to the general meeting; and if the petition is favorably regarded—as it usually is—an appropriation is made for the reading-room, which is ordinarily placed over a branch store. These reading-rooms have attached to them rooms for conversation and recreation, and it was here the co-operative cotton-spinning mill was started.

In these rooms during the year 1885, the subject of discussion, night after night, a little History of the "Self-Help Society" tells us, was: Cannot some scheme or method be adopted whereby it might be possible for the working classes of Burnley to become their own employers? After the subject had held the field for weeks, and every side had been shown up by its advocates or opponents, it was decided to try a cotton-spinning manufactory. What they aimed at, the History says, was to reverse the joint-stock system by making capital the servant of labor, and paying it only the market rate of interest, not allowing it to absorb all the profits.

"The bulk of the men," says the History, "who attended the discussion had spent a life in the cotton trade, and were quite alive to the fact that if they were to start to manufacture cotton goods, they must have a mill equally fitted with any of their competitors; it must be furnished with the most modern of machinery, and must be able to start in such a way as to utilize all its machinery to the best advantage.

"This made the venture a big one for a lot of workingmen to undertake; the lowest sum we could manage to start with, and start in a way to give us a chance of being successful, being fixed at $25,000. This meant for machinery and trading capital only. The mill: we proposed not to build one, that was beyond us, but to rent one on a system which is very common in Northeast Lancashire, where you pay so much per annum for room and power for each spindle and loom. This enabled us to start in rather a large way, for, remember, we could not depend on anybody taking shares only workingmen and co-operators.

"The traders and manufacturers, as a rule, all ridiculed our efforts, the active members of the society being treated to expressions of this character: 'Well, which of you is going to be master yonder, or is it going to be everybody master?' But they managed to survive all that.

"There were many anxious nights spent by the provisional committee that had been appointed, in studying out ways and means. They canvassed their friends, appeals were made to surrounding co-operative societies, with no large measure of success. When the thing looked hopeless to find all this capital amongst workingmen, a gentleman came forward with an offer of a loan, which enabled the committee to see their way to go on with the scheme. A mill was taken on a lease for fifteen years, with an option clause of five years longer at a very easy rent; in fact, as low a rent as was paid for any similar convenience by any firm in the district. In their first half year's balance-sheet they were able to declare a very handsome profit; up to this time the directors had received no salary, in fact, the whole of the formation expenses, which included cost of deeds and legal expenses of $150, was but $335. This was cleared off their first balance-sheet, and after providing for interest on all share and loan capital, also depreciation charges, a bonus of five per cent was paid to labor, and over $2,000 was carried to a wage-balance fund. Just a word here on the subject of the wage-balance fund:

"It had been acknowledged on all hands during the discussion of how to carry on successfully a productive society, that there would always be a danger in asking any body of workpeople to submit to paying their share of any losses, but that this might be met by creating a wage-balance fund in prosperous periods, to tide over periods of depression, when the concern was liable to make losses. The system they adopted was to pay part bonus, and the other portion which was carried to wage-balance fund, was, when it amounted to $5,000, to be allotted out to the workers in the same way, and ascertained by the same basis as all other profits."

Thus from the beginning these workers, in arranging to be shareholders, had provided that if profits did not admit of the payment of the interest on share capital, and the fixed charges for depreciation of machinery and land and buildings, each worker should pay his share of the deficit, calculated on the same basis as that on which the profits had been distributed to them.

Unhappily, it was not long before this wage-balance fund had to be drawn upon. After two years of very prosperous business, the society had to call upon the workers to pay losses. In twelve months the working people paid back over $4,000. Every one paid his share, and only with about a dozen was there any difficulty in getting them to face the responsibility. This was followed by another period of prosperity, under a new manager, who still remains. The wage-balance fund was built up again in a year, to the sum originally fixed, of $5,000, requiring about $75 from each man, and about

$50 from each woman employed. On this they receive interest at the rate of four per cent per annum.

The workers hold over two-thirds of the capital. The Burnley balance-sheet, in giving the usual statement of the attendance at the committee meetings, adds to each name a word of description of each officer or committeeman. From this we learn that of the seven committeemen, the chairman and three others are workers in the mills; one is a non-worker (representing, no doubt, the outside individual shareholders), and the two others represent co-operative societies which hold stock.

When I was at Burnley, the society was again struggling with very hard times, as was the entire cotton-spinning industry of Lancashire. Mr. Jones, in his book, "Co-operative Production," implies that charges of bad faith have been made against the present manager of the Self-Help society. Although he does not say so, his text prepares the reader to believe that its embarrassments are largely due thereto. I heard nothing of this, and there was no suggestion of the slightest want of confidence in the management, on the part of employés, or on the part of the Labor Association and Productive Federation, which, of course, make it part of their business to keep themselves thoroughly posted with regard to the condition of such establishments. In fact, after these charges were promulgated, the employés passed a vote of confidence in the manager, requesting the directors to make a public denial of the charges against him. The only criticism I heard from other co-operators was that the society had weakened itself by not making sufficiently large additions to its reserve fund during its early prosperity; but the facts just recited with regard to the wage-balance fund would seem to justify the statement that the precautions taken in times of prosperity to prepare for adversity were at least equal to what is customary in private enterprise.

The business of the co-operative store at Burnley I found to be decreasing heavily, on account of the hard times in the cotton industry. When I was there, 222 looms in the mill were idle out of 1,034. Meanwhile, the Co-operative Wholesale Society of Manchester, and other co-operative stores, were buying cotton cloth largely of other than co-operative mills. Two-thirds of the Wholesale's purchases of cotton goods, I was told, are made of capitalistic manufacturers. It is easily conceivable that one of the causes of the troubles of the Burnley mills was this failure of the co-operative stores, which should be their most steady customers, to give them support.

I found that the society was in debt $10,000 to the bank, and that the seven directors had given their personal bond for this amount; but they did not then feel that they were in any danger. There was certainly every evidence of the most careful and economical management. Every nook and corner of their land was in use. Even a little angle of waste space had been rented for a woodwork shop, where pegs were made for bobbins. There was a smaller proportion of idle looms, they told me, in the Burnley Self-Help mills than in the average of the trade. They made a larger variety of goods than others, and had over $100,000 worth of home trade on the steadiness of which they could rely; while the private manufacturers have to sell in the open market on the Manchester Exchange. Another advantage here—as everywhere throughout Co-operation land—is that there are practically no bad debts, as the sales are all made on very short time and in the home trade, to co-operative societies which almost never fail to pay.

The sacrifices which the Burnley Self-Help weavers made in bearing the burden of losses have been so bravely met that it is to be hoped the anticipations of its friends will be realized, and it will be able to continue in business. One of the criticisms most strenuously leveled against the Burnley works is the contract which has been made with the present manager. He was their manager in their early history, and was compelled to leave by dissensions among the members, who would not pay him a proper salary. When crowded out of Burnley, he went into business on his own account and made a fortune. But the co-operative work had so strong a hold on him that he afterwards went back to the management of the Self-Help society. After the disaster which followed his withdrawal, and the first exhaustion of the wage-balance fund, the society recalled him. As he had a large business of his own to manage, he was unwilling to return without some assurances of permanence; and he therefore stipulated that he should be appointed as manager for ten years, and that he should be subject to no undue interference by the directors. In case of a dispute between him and the working people, an arbitrator was to be called in. He pledged himself to run the mills at a profit, or to do without wages; that is, his salary was to be half the profits made on the sales in the open market, the other half to be paid to the workers as a bonus. He was to have no claim on any profits arising from the home trade, which were to be divided between the workers and the consumers. It was also agreed that if, under his management, the society made losses and the wage-balance fund became exhausted, the agreement could be cancelled by a meeting of the members at any time.

The high esteem in which the Burnley Self-Help workers were held in their town was shown by a legacy of $2,500 left them last year by one of the manufacturers of the city. This was given as the nucleus of an old age or superannuation fund; the members of the factory were left to arrange the details of the scheme. The gentleman who made this thoughtful bequest was a private manufacturer, Mr. John Williams, who had risen from the ranks of labor to great business success, but always felt and acted in sympathy with the efforts of working people to better their conditions by co-operative action. In addition to this gift, he left $3,000 to the workers in his own mill, and other sums to various public purposes.

The productive committee of the Co-operative Union have passed a resolution, after hearing the statements of the auditor and liquidator, that they consider the society to be sound if it can get capital enough, and recommend it to the support of the co-operative movement. One of the workers in the Burnley mill said to me: "Co-operation cannot be stopped." It is to be hoped these words are as much a prophecy of the future of the Burnley Self-Help as of the movement at large.

Labor Copartnership for May, 1898, says with regard to the proposed scheme of reconstruction of the Burnley mills: "This society went into voluntary liquidation, not because it was insolvent, but because capital did not come in freely enough to finish the extensions of premises undertaken. Voluntary liquidation made it possible to protect the interests of everyone concerned. For the reconstruction about $55,000 will be required, and this it is proposed to raise on mortgage bonds at 4 per cent., redeemable at the expiration of ten years or less. The bonds will be a second charge (after the first mortgage) on all the freehold premises. The circular proposing this plan is signed by the liquidators and on behalf of the Productive Committee of the Co-operative Union."

The operatives themselves are taking all they can of these bonds, the London *Daily Chronicle* of May 6th notes, and have agreed to pledge their wages as additional security for those sold to outsiders.

CHAPTER IX

MUSHROOMS THIRTY YEARS OLD

At Hebden Bridge, not far from Burnley, and at Paisley, in Scotland, are two of the oldest copartnership works in the movement—"mushrooms" twenty-eight and thirty-six years old, and still growing. The Paisley Co-operative Manufacturing Society, founded in 1862, like the co-operative works at Kettering was a direct outgrowth from the co-operative store. Co-operation found Paisley ready for it. The place has always been famed for its political fervor, and in the old times was full of radicals, chartists, and poets of the people. The weavers, sitting around their looms, were continually talking about employing themselves, and out of this came the store, and then the factory. The Paisley works, which manufacture woolen and cotton goods of a great variety, have now been in successful operation thirty-six years. They made the same humble beginning as all the co-operative societies, and for some time the house of their secretary was not only their committee-room, but also the warehouse where their goods were stored.

They began with seven members and a few shillings, and now have a membership of 262 societies and 1,346 individuals. The practice of giving a share of the profits to all the workers of the society was adopted in 1869, one year after the adoption of the plan of sharing profits with purchasers. Their balance-sheet for the second half year of 1896 shows their capital to amount to $284,185; the sales for the year to $360,095; the profits, $16,670, after making the usual allotment to the insurance and reserve funds; there was $8,715 of profit for the half year for dividends to customers and working people. Of this, the customers received $7,375 on purchases by members amounting to $177,000, and the workingmen $1,340 on their wages of $32,165. The society owns its own land and buildings.

This is one of the few copartnership establishments which have had trouble with employés. There was a strike in 1891, and for some time the society employed non-unionists; but the dispute was settled, and the society agreed to use its influence to have its workers join the union. In 1892 it was proposed to take away from the employés of the society their right to serve upon the committee of management, and a majority voted for the proposition; but, as it was not the two-thirds majority which was required to alter the rule, the motion was lost.* The workers are all members of the union, and all receive the highest wages in addition to their bonus. The weaving shed is of the most modern construction, well ventilated and lighted. When I was there, the society was talking of putting up more buildings. In the last six months of 1897 the sales were $187,077, an increase of $3,630 over the corresponding period of the year before, in the face of the great depression caused by the Engineers' strike and the bad condition of the cotton trade. The sales for the whole year were $370,705, as against $360,095 for 1896. The profit was $17,465. A dividend of ninepence in the pound was made on purchases and wages. Out of 328 employés, seventy-one are shareholders, but only to a small amount.

Just before we were at Paisley, the organization of a co-operative laundry had been begun as the result of a boycott. The incident is one of a hundred going to show how the war of the private traders has strengthened and widened the foundations of the co-operative movement. A biscuit manufacturer at Barhead had thrown out of employment a number of girls because

* Co-operative Production. Benjamin Jones. p. 342.

their parents or other members of their families belonged to the co-operative store. Mr. James Deans, who, as representative in Scotland of the Co-operative Union, has such work as this to do, was called to attend a meeting to find some work for the disemployed. Of the many projects discussed, a laundry was finally decided upon. It was certain of support, as it could get work from the Baking Society of Glasgow, the Scottish Wholesale, and the Paisley manufactory. The sum needed was $7,500. When I was there they had already secured $5,000 of it. Plans had been made, and the building was to be put up immediately. This laundry was to be run on copartnership lines. It has been the boast of the Scotch co-operators that in Scotland, every one of the productive societies—which, however, are not numerous—shares profits with the workingmen. But at a meeting to organize the Co-operative Laundry Association in Glasgow, February 19, 1898, the principle of paying a bonus on wages was rejected by a vote of 21 against 18.

The Hebden Bridge Fustian works are among the oldest of the copartnership enterprises, and gave the first rays of clear light after the darkness that followed the failure of the workshops of the Christian Socialists. Mr. George Jacob Holyoake says: "The beacon fire of courage and hope was lighted on the hills of Hebden Bridge, and for more than a quarter of a century it has been a signal flame of unity, sagacity, growth and profit." Mr. Thomas Hughes said before the Co-operative Congress of 1887: "It has never had a strike, or even a serious dispute between managers and working people; and its effect and influence on the neighborhood, both economically and morally, has been as remarkable as it has been admirable." Its steam engine bears this inscription:

"STARTED BY
"E. VANSITTART NEALE, M.A.,
"NAMED
"THOMAS HUGHES
"BY THE MOST HONOURABLE
"THE MARQUIS OF RIPON, K.G.,
"OCTOBER 5, 1887."

This Hebden Bridge enterprise, like those of Paisley and Burnley—and, in fact, practically all the productive co-operations—was the lineal descendant of a co-operative store in its vicinity. This store was founded in 1848, and was one of the first to carry forward the example of the Rochdale Pioneers. Hebden Bridge is especially attractive to the co-operative traveler for the picturesque beauty of Nutclough Glen, owned by the works, with its rocks and trees, and

THE PICTURESQUE BEAUTY OF NUTCLOUGH GLEN

Owned by the Works

the cascades from which the mill receives the soft, pure water that is indispensable for its manufacture; and attractive also for the presence there of Joseph Greenwood, who was one of the founders of the enterprise and who is, and has been from the beginning, one of the "Old Guard," faithful to the ideal of the rights of labor to a voice and a share in its own production. Mr. Vansittart Neale, after the failure of the Christian Socialist workshops, devoted himself largely to the organization of the annual co-operative congresses. The first of these was held in London in 1869, and the second at Manchester in 1870. This congress was attended by Mr. Greenwood and one of his associates, and gave them the immediate inspiration to enter upon co-operative manufacture. The trades-union movement was strongly represented in this congress, and one of the points on which most emphasis was laid was the folly of the trades-unionists' policy of accumulating large amounts of money and using none of it in productive works to give self-employment. Mr. Greenwood was a member of the co-operative store, and his fellow-delegate belonged to the Fustian Cutters' trades-union. The discussions of the congress made a deep impression upon their minds and suggested, what had not occurred to them before, that it was possible to start a productive workshop.

A tragic event—one of the classic incidents of co-operative literature—which occurred just after this caused them to hasten in their new plans. There was among their fellow-workers an old Irishman, seventy years of age, who had lived in the midst of the eventful period of the agitation for the repeal of the union with Great Britain, and who could tell, as an eye-witness, the story of the massacre of Peterloo. He was too feeble to carry his own piece of cloth from one part of the works to another, a quarter of a mile away; so his comrades took turns in doing this for him. On one occasion a dispute arose as to whose turn it was. Meanwhile the old man went off himself for the load, and when he brought it in, sat down and died. He had to be buried by the contributions of his mates. The funeral was a keen reminder to the rest of the necessity for combining, if they were going to survive the struggle which provided only the bare necessaries of life, and was always open to the possibility of their being unable to stave off the idleness which would bring poverty, hunger, sickness, and premature death. From this outlook to the thought of starting a burial and friendly society was but a step, and this, what they had heard at the co-operative congress, made it easy for them to carry one step farther into co-operative work.

Thirty of them began making contributions of threepence a week for this burial and friendly society, and for a fund to be used towards setting up a fustian cutting and dyeing establishment, with the determination to find employment under their own control. These threepences a week of thirty workingmen in 1870, had by 1898 become a capital of $188,335, with a trade of $224,200. By the end of 1896 there was a membership of 314 co-operative societies, 309 workers, and 199 outside shareholders. The amount of profits paid during that period amounted to $318,705; $900,085 had been paid in wages, and the bonus paid to labor had been $32,580. A reserve fund had been created of $19,685, an insurance fund of $20,130, and there had been written off for depreciation $63,-815. The employés hold $38,685, or twenty-nine per cent of the capital. The report for the half year ending December 31, 1897, after allowing $2,745 for depreciation, shows a profit to be divided of $11,425, of which labor receives $1,925; $200 is carried to the educational fund, $500 to the reserve, and $750 to the insurance fund. The sales for the half year were $116,705.

The story of the means by which these results were achieved is a human document of no mean interest, and has been related by Mr. Joseph Greenwood, the manager of the society, in a little pamphlet, "The Story of the Formation of the Hebden Bridge Fustian Manufacturing Society." "We were poor men," says Mr. Greenwood; "with the exception of one or two of our number, none of us owned as much as a five-pound note. We, nevertheless, put our threepences together week after week, until we got about $50, which we invested at interest" in the Hebden Bridge co-operative store. The sum which they needed to buy a dyeing plant was $5,000, and the prospect of obtaining it seemed remote enough. "We calculated and schemed," says Mr. Greenwood, "how we could make progress and increase our little capital. We rented a small upper room over a passage, ten feet by ten feet." . . . . "Our spare time was given to fitting up the small fixtures we could afford, buying the boards and making the shelves ourselves, which served to hold our small stock. We bought a second-hand chair and table, and with two forms we were completely furnished for a meeting-room." Here they held their meetings, trudging back a mile and a half to their homes at eleven o'clock every night in the week.

HEBDEN BRIDGE FUSTIAN WORKS
"We put our threepences together week after week"

They bought four pieces of cloth to cut—there are thirty-six hundred threads to be cut by hand in every square inch by the fustian cutter—and were very much encouraged when they were able to sell two or three half-pieces to the neighboring stores. Stores that were near enough they visited in the evenings after their day's work at fustian cutting had been done; to those that could not be reached by walking and returning the same evening, they sent patterns of their goods by mail. It was a very uphill fight. Singularly enough, their expectation that the co-operative stores would be their best customers was not at first verified. Some of these seemed to think "that we wanted trade simply because we were co-operators," and would not even look to see whether the things offered were good or not. "We had thought that we were going to have only one difficulty, viz., want of capital; but we soon found that to obtain trade was equally difficult."

At the beginning, none of the founders allowed themselves any pay. All the office, warehouse, and other work for two years was done for love. At the end of their first half year they had spent $32 for expenses, including a complete set of account books, and had in stock on hand $397; had done a business of $275, and made a profit of $15, which was carried forward. At the end of the first year, they had to go into larger premises. "The society grew," Mr. Greenwood says; "it was fostered and lived in us and formed part of our lives, and was always in our thoughts. No day was too sacred for us on which to devise and seek the association of men whose bent was similar to our own, and confer with them on our and similar undertakings. It may shock some minds to know that we met on Sundays and talked about our obstacles, and how we made progress. At first, our social Sunday gatherings were confined to the nooks and corners of the valley of the Hebden and at the firesides of friends near. Then we planned excursions further. Friends came to see us, and we went to see them. The fields and the lanes, the wild flowers and the ferns, just opening their young fronds, and the bright green tints of the tender leaves of spring were to us made more beautiful. The purple moorlands and the nut-brown shades of October had charms made all the sweeter and mellower. The streams from the hills dashing down their stony beds, the glistening of the sunlight on the white roads and on the mossy and fern-tufted banks by the river and the footpath, were made more delightful. We felt we were doing God's work, and in that faith and communion we were content."

By the middle of 1873, the workshop was going on regularly; employment was constant, and the members of the society felt themselves "comparatively removed from anxiety and doubt" in regard to their earnings. Among other causes which contributed to this success was the fact that the society had directors who directed. As an instance, the half yearly balance-sheet of June 30, 1897, gives the following as the attendance of the managing committee:

ATTENDANCE OF COMMITTEE.

	Expected.	Actual.
Joseph Craven, President,	24	24
James Johnson,	24	22*
Arthur Ainley,	24	24
Thos. Wadsworth,	24	24
John Harwood,	24	24
Thos. Hy. Pickles,	24	24
Wm. Hy. Helliwell,	24	24
John Waddington,	24	24
John Tootill,	24	24
Abram Haigh,	24	24

*Twice sick.

The publication of such a list of attendance at committee meetings is a usual feature of co-operative balance-sheets.

The idea of joint interests and ownership was constantly before these men from the beginning. It was with the hope of carrying out this plan, to which Owen and the Rochdale Pioneers and the Christian Socialists had been devoted, that these fustian cutters of Hebden Bridge had embarked in the formation of their society. At the end of the second half year, the managing committee recommended a meeting of shareholders to give a dividend to labor at the rate of five per cent. There was "a stiff opposition from some friends whom we had taken into membership who were not with us when we started. This idea of a dividend to labor created some excitement, too, among the outside public; and although a small part only was recommended, it was regarded as a new-fangled notion, and one that could not be just to the shareholder." A dividend to labor in the Hebden society preceded a dividend to the purchaser, reversing the order of movement in the Paisley society.

The society was only a few years old when it was able to pay dividends at the rate of ten and even twelve and one-half per cent. "The joint-stock movement was then in high favor. The spirit of little capitalists was insatiable, and even co-operators, so-called, hastened to get rich in this fashion. We felt we were under the necessity of checking this spirit." This was done by reducing the rate of dividend on shares to seven and one-half per cent. Later, the rate of dividend was reduced, after a severe struggle in which feeling ran high, to five per cent. The outside shareholders had come to regard their rate of interest as a vested right, and denounced the proposed reduction as a confiscation. They made no allowance for the fact that during the years they had been holding their investments they had been almost twice repaid, and still had their shares. It required a prolonged agitation, in which several defeats were encountered, to get the two-thirds majority required for this reform. The practical effect of this reduction of dividends to capital has been that the money thereby saved was available for the extension of the trade of the society. The practice of a high rate of dividend on share capital, Mr. Greenwood thinks, is one of the greatest failings in the co-operative movement. The co-operative stores have paid five per cent on money for which they could not get more than two or two and one-half elsewhere; and in this way they have punished their trade and the poorer purchasing members, and have been unable to supply funds for productive enterprises unless at high dividends.

Another evil came from their prosperity to embarrass the society in its efforts to realize its ideal of promoting the interests of the workingmen as its chief purpose. This ideal had been expressed in the first rules of the society as follows:

"The object of the society shall be to find employment for its members" . . . and to "practically educate its members in the causes which operate for and against them in their daily employment, and in the principles that will tend to their elevation and improvement."

Employés had been left free to sell their shares upon leaving the society. It had shut its share-list to outsiders, but its rules allowed any shareholder who had not taken the full number of shares (100) to do so. The high dividends which were being paid upon these shares had carried them to a premium, and outside speculators began buying up the one or two shares owned by employés who were leaving the society, and taking advantage of the right which went with this ownership of subscribing for enough more to make up the holding of one hundred shares. This was bringing in outsiders in a constantly increasing number. The society began to face the danger of seeing all the objects for which its founders cared most completely defeated, as had happened in the mill of the Rochdale Pioneers and the first workshops of the English Co-operative Wholesale. These outsiders got control of the committee which had in charge the revision of the rules. They were opposed to labor having any share in the profits whatever, and set themselves to take it away, though it was the very heart of the society, by submitting new rules, omitting all provision for a share of the profits to the workingmen. Mr. Greenwood gives an account of the meeting in which this crisis was settled.

"It looked," he says, "as if the doom of our most vital principle had come. The business of the society was going on all right, but disagreements and divisions prevailed on the committee and among the members. It was a serious time. The general meeting was held to consider the altered rules. The proposed alterations had become generally known in the co-operative movement, and one of the leaders, Mr. Greening, wrote a letter to the co-operative *News*, calling on all shareholding societies to give their support on the side of sharing profits with the worker, and during the meeting a telegram arrived and announced that the Scottish Wholesale supported the principle. A few individual members were in favor of the workers, and, mostly, the society representatives were on the same side. Minor alterations were agreed to, but the signs of strife were upon the countenances of the chief persons who were about to take part in the main struggle. The chairman seemed to be in a nervous state. The society's representative, who, above all others, should have been in favor of the worker, led the way, and moved the adoption of the altered rules. This was seconded. There was one member who was in at the foundation of the society who sat gloomy, drawing his breath very quick, but feeling intensely moved. He was not gifted with the power of speech, but he felt the duty lay upon him, and he got up and clearly stated the purpose for which the society was started, going from point to point, and charging the promoters of the alterations with breaking an obligation and doing a great wrong to the men who were the founders, and who had made the society a success. He moved the rejection of the altered rules. This was seconded by a former president and one who had done great service to the society. He could not let go the principle for which, as co-operators, they had striven so much, and felt bound to support it. The meeting afterwards drifted into a most inextricable confusion, when a delegate from a large society in a Yorkshire town got up in the midst of the hubbub, whistling a popular tune. The audacity of the act at once arrested the attention of the meeting, and before the chairman had time to call order he at once addressed himself to the question at issue, giving his support to the latter motion. There could be no doubt now about the result, the meeting was so demonstratively in favor of the speaker's argument. The altered rules were rejected. Mr. E. Vansittart Neale afterwards made the alterations to meet our requirements. These were adopted and registered. The society was saved. It was destined to do still more for the worker. It was to be a leader in the van of the organization of labor."

In the year 1873, while engaged in these struggles to cut down the share of the capitalist and to preserve the share of the workingman, the members of the society had such undisturbed confidence in its future that they entered upon the most important enterprise of its history—the purchase of the Nutclough Glen and mills, in which they are now settled. For the trade they were then doing in fustian cutting the purchase would have been a great folly; but with the dyeing trade that they could do with these new premises, with its stream of soft water, there was a good prospect of success. The society had no money of its own for so great a venture, and obtained it by a loan on mortgage from the English Co-operative Wholesale Society at Manchester. The cost of the estate was $27,250, and the alterations and new machinery cost $17,570 more.

The Hebden Bridge society is among those which have incurred the reproach of not allowing workingmen to be eligible for places upon the board of direction; but, following the progressive policy which has always characterized it, its rules have been revised recently so as to recognize this principle, and it is now, in all respects, a labor copartnership society. The society, it will be noticed, is not a "self-governing workshop." Among its 822 members are 314 co-operative societies and 199 individual shareholders; all of whom have a voice as well as the 309 workers.

An interesting table, published in one of its pamphlets, gives the statistics of some personal instances of the benefits of the society to its working members. The cases of six men and four women are tabulated. Their occupations were: Garment finishers, machinists, fustian cutters, laborers in the dye-house, and weavers. Their length of employment ranged from nine to twenty-four and one-half years, averaging seventeen years. They had received a bonus on wages ranging from $77 to $233 each, and averaging $142. The amount they had paid in ranged from nothing to $500 each, an average of $186. The amount withdrawn varied from $18 to $500, an average of $224, and each still has shares ranging from $70 to $500 each, and averaging $223.

As we walked through the works, it was pleasant to note, as one does everywhere throughout the world of co-operation, the friendly relations between the managers and the members. Stopping before one of the fustian cutters, Mr. Greenwood said: "Here is an old socialist and co-operator," with a friendly laugh. "You think," I said to the man, "that socialism and co-operation are not irreconcilable?" "No," he replied; "perfectly reconcilable." We stopped before a room full of women, all full of happy interest in their work, and all exchanging cordial glances with the head of the establishment. They were all part owners in the factory, all shared in the profits, and all had a vote in the management, unless they were under sixteen. In that case, their share of the profit was accumulated to pay for the stock allotted to them. "At our half-yearly meetings and conferences," said Mr. Greenwood, "the men and women have always taken an interest in the business, giving what support their judgment and ability dictated."

The Yorkshire *Factory Times* of March 5 and 12, 1897, comments upon the condition of these working people as follows: "No one can fail to be struck with the difference in appearance between the operatives at Nutclough and at similar firms. Their labor price is higher than that given by private firms which compete with Nutclough, although not as high as in some parts of Lancashire. The material comfort of the workers, many of whom come down from the hillsides to their daily toil, is well looked after, sanitation and cleanliness being a marked feature. The 'dining-room' is replete with the necessary utensils and accommodation, including one of the highest class gas stoves for cooking purposes. Whilst not without married women, the number of wives with families is comparatively small. The machinery room gives more air space, by far, than is required by law. There were a few men (cutters) engaged among the crowd of feminine beauty, who seemed to enjoy their lives as heartily as the women. Among the three hundred and thirty workers at Nutclough, there are only five half-timers. The women, in contrast to some ready-made clothing firms, are not forced to buy their sewing materials from the firm. The Nutclough directors serve out to their women every necessary commodity to finish the article the women are engaged upon."

Almost none of the working people are trades-unionists. The trades-union is very weak all through Yorkshire, because, it was explained to me, the working people are too content. The attitude of the outside workingmen towards those in these works is described as one of indifference with a little jealousy.

The democratic relations between the workingmen and the management, and the fact that only picked men and women are likely to be found in co-operative industry, are an important factor in its political economy, and count for a great deal in settling the question which is the fittest that survives — co-operative or competitive industry. For one thing, the co-operative workers save a good deal in the cost of superintendence. They also get a better quality of work. "Our advantage over our competitors in marketing our dyed goods," said the manager, "is simply through the better quality of our work." They take pains to keep abreast with the improvements in manufacture. One of the Hebden Bridge dyers was sent to the technical school at Huddersfield, with the result of a saving of fifty per cent in the cost of chemicals. The workingmen have nearly $40,000 of the stock of the factory, and have more in the store. These resources qualify them to stand twelve months of idleness without breaking down. "Every co-operative workingman in this place," said Mr. Greenwood, "has $250 to the good, and in most cases sons and daughters who have more."

I never failed to be stirred afresh by every evidence that I saw of the high ethical and sympathetic point of view which was taken by the co-operators, and which, to one familiarized with the low tone of private business, is as strange as it is pleasant. Passages like the following are constantly cropping out in the prospectuses and advertising pamphlets of these "business" enterprises. In "A Brief Sketch of Twenty-six Years' Work in Co-operative Production," issued by the Hebden Bridge society, it says: "The position of employer is an important trust in regard to remuneration of labor, in regard to the provision of educational facilities, and in regard to social relations, both in reference to the workshop and citizenship. In these respects the society has done something, and there can be no doubt that in the homes of the workers there has been a larger measure of comfort and a higher standard of rational enjoyment. The working members of the society do a fair share of service on the committee of the local store, on the educational committee, and also in connection with other public institutions —political, social, and educational."

These associations of workingmen regard their business concerns not merely as commercial enterprises, but as social and moral agencies. The pamphlet just quoted from continues: "At present the whole range of industry is carried on and controlled by a grand army of irresponsible employers, too much engrossed in their own aggrandizement. We shall never get the full fruits of labor for the laborer until there is all around an intelligent appreciation of the merits of true co-operative production."

In examining the cash accounts of the Hebden Bridge society, I find, as almost universally in co-operative accounts, numbers of expenditures for charitable purposes. There are here subscriptions to the Royal Albert Asylum, the Halifax Infirmary, and also to the Halifax Infirmary Building Fund, to the Devonshire Hospital, the Deaf and Dumb Institute, to the Penrhyn Quarrymen (then on a strike), to the Cowdenheath Disaster Fund, and to the India Famine Fund, besides the usual appropriation to its own educational fund.

I asked the leaders here what was their expectation as regards the ultimate development of co-operation. They believe, they told me, that the co-operative system will slowly displace the competitive, and look upon all the temperance, trades-union, political and socialistic agitations as merely preliminary, preparing the people for higher forms of association. They think the government will come to own many more of the common industries; and our business men giving up their present privilege of being industrial buccaneers, will submit to the social discipline which has given us policemen and soldiers under the authority of the state in place of the lords and barons of the middle ages waging private war. The private manufacturers "cold-shoulder" co-operators much less than formerly when they meet them now at district councils and upon other public bodies, though they still show sometimes a little of the old spirit. They begin now to see what co-operation means with regard to their economic future, but they also see that it is too late for them to stop it.

CHAPTER X

WORKMEN DIRECTORS IN A GREAT COMPANY

There is in the labor copartnership world something more novel even than the associations of workingmen and workingwomen owning and managing, sharing profits and losses. The South Metropolitan Gas Company of London, with a capital of $35,000,000 and receipts of $3,300,000, employing almost three thousand men, shares profits with them, makes them shareholders, and this year these workmen shareholders will elect two of themselves directors of the company. This admission of the workingmen to copartnership was the result of the organization in 1889, following the dockers' strike, of a gas-workers' union, and of the conflicts which then arose between it and the company.

But the history of this company as an employer—which is really the history of two men, the present manager, Mr. George Livesey, and his father, Mr. Thomas Livesey, who preceded him—shows it to be a continuation of a policy dating back a generation. On the recommendation of Thomas Livesey, this company gave its men, twenty-five years ago, an annual week's holiday with pay. Some years later the holiday with pay was doubled to employés of three years' standing, and is, therefore, now practically a fortnight's holiday with pay. Mr. Thomas Livesey also started sick and superannuation funds. In case of sickness, twelve shillings a week is paid and medical attendance given for thirteen weeks; and where it is needed, it is the custom to extend this for three months more. When the workingmen become unable to do a fair day's work, they are given pensions of ten shillings to sixteen shillings a week, according to length of service. Each employé is required to contribute threepence a week to each of these funds, but this does not provide enough for the benefits, and they are maintained at a heavy cost in money to the stockholders. The company was active in the abolition of Sunday day labor in gas works, and gave its men this exemption without solicitation from them. It also took the initiative, in 1887 and 1888, in proposing the change from the twelve-hours shift to the eight-hours shift, but the men did not meet it. When, after the organization of their new union, in 1889, they asked for the change to eight hours, it was conceded, not only without reduction in wages, but with an increase estimated to be twenty-five per cent.

In 1876 this company was put by Parliament, as to the prices it was entitled to charge for gas, on what is known as the sliding-scale basis. The standard price of gas being taken to be three shillings sixpence a thousand feet, it was allowed to pay a standard dividend of four per cent as long as it kept its price down to that figure; but for every reduction of one penny below this price, it was allowed to increase its dividend by one-quarter of one per cent. Mr. George Livesey objected to this arrangement then as incomplete, because it left out the workingmen. He desired that the sliding scale should be extended so that their wages also, as well as the dividend of the stockholders, should increase as the price of gas went down; but he was unable then to remedy this defect in the scheme. Mr. Livesey always had profit-sharing in mind, and in 1886 he succeeded in getting the consent of the directors to a scheme to give it to the officers and the foremen, though he was still unable to get his associates to include the other workers.

The Gas-Workers' Union, formed in 1889, set to work to enroll all the men employed in gas-making in Great Britain, and entered upon a forward policy, inspired by the elation which had taken possession of the laboring men in consequence of the brilliant success of John Burns, Ben Tillett, and Tom Mann, in the cause of the dockers. It made demands, some of which were granted; and when others were refused in one place or another, it struck. In some cases the management lost control; in others, the strikers were defeated, and all through the industry a feeling of great uncertainty and uneasiness prevailed. Mr. Livesey was one day notified by the foreman of his works that, owing to the influence of the new union, the men were getting out of hand. The yardmen would soon be forced into the union, which would then become the master of the establishment. To head off the union, the company reduced the hours to the yardmen and paid them higher wages. But it was plain that this was not enough.

Mr. Livesey then decided it to be better policy to spend money to satisfy his workers than to fight them. He therefore proposed to the directors that they share profits. "We want more than the workman's labor; we want his interest," he said. The men were to share profits upon signing agreements to serve the company for twelve months. To make this profit-sharing more attractive, it was dated back three years, so that the men should receive from the profits of the past "a nest-egg," to which they might add their profits for the future. This nest-egg was equal to eight per cent on the year's wages of those who had been with the company three years. This was to remain at interest for three or five years, except in case of death, superannuation, or leaving the service in an honorable manner.

The agreement to work for twelve months was not designed to prevent the men from leaving the service of the company, but to prevent their leaving together. It was provided that any man could leave by consent of the engineer; and as there were always men applying for positions, and as any man can be spared from any place, the men could practically go when they wanted, but they could not leave in a body. The agreements were arranged to terminate on different dates. Certain cases had occurred in other gas works in which the men had started to leave their work all together with no notice whatever, thereby compelling the company to accept in every particular all the terms they had demanded. It was to prevent such a surprise, and also to make impossible the disastrous contingency that a large part of London might be left in darkness, that these agreements were proposed. When this bonus and agreement were offered to the men, the Gas Workers' Union directed its members—the stokers—not to accept them. But the unorganized gas-workers said to the company: There is no reason why we should have none of the benefits of this agreement, even if the trades-unionists do refuse it. The agreement was thereupon offered to the men individually. "If only one man," said Mr. Livesey, "wants it, we will accept it from him." A thousand men—all the non-unionists—signed in fourteen days.

This profit-sharing scheme of 1889 was arranged in this way, realizing the plan which Mr. Livesey had ineffectually sought to carry out in 1876: For every penny by which the price of gas was reduced below two shillings eightpence per thousand, a bonus of one per cent on wages and salaries was to be paid annually; and, as already said, this bonus scheme was dated back three years. If the workingmen chose to leave their share of the profits on deposit with the company, they were paid four per cent interest upon it. Another concession made to the men was that they were allowed to have gas on the same terms as the directors—at cost price. From the beginning the workingmen left thus on deposit nearly one-half their bonus, besides depositing other large amounts of their savings, and also investing considerable sums in the stock of the company. Some of the stokers signed the agreement on the sly, violating the orders of their union. The union demanded the dismissal of these "blacklegs" and the abandonment of the bonus and agreement. This was refused, and the well-remembered gas-workers' strike of London, from December 13, 1889, to February 4, 1890, followed, ending in complete failure.

The bonus scheme did not at first exclude the trades-unionists. In an interview with the workmen of the company, just before the strike, Mr. Livesey said: "I have carefully avoided all reference to unions. We do not wish to interfere with the union in any way. We make no restriction about it. If a man likes to continue in the union after signing this agreement, he can do so. We ask no questions." "I have told the men," said one of them, "to stick to their union." "By all means," Mr. Livesey replied, "so long as they do what is right." But within two weeks after the settlement of the strike, and while the company had a number of union men in its employ, the secretary of the Gas Workers' Union said in a public speech, referring to the recent trouble: "The men will not give seven days' notice again before striking." Taking this as a declaration of war, not only upon the company, but upon London, Mr. Livesey, for the first time, adopted the policy that every man who signed the profit-sharing scheme should declare that he was not a member of either the Gas Workers' Union or the Coal Porters' Union, which had acted with it during the strike. And in his public utterances, Mr. Livesey now declares his system to be "quite incompatible with trades-unionism. The two things are diametrically opposed. Trades-unions are the union vs. the employer; in our case it is the union with the employer, not against him." "Union of capital and labor is better than a union of laborers and a union of capitalists fighting each other," is one of Mr. Livesey's ways of putting his position. A copy of the contract offered the men was given me by Mr. Livesey. Its first section reads: "The said South Metropolitan Gas Company agrees to employ the said . . . . , who says that he is not a member of the Gas Workers' Union." But Mr. Livesey, when before the Labor Commission in May, 1892, said that membership of the unions by his men was being winked at as long as the members kept quiet. *Labor Copartnership* of November, 1894, commenting upon the statement in a letter in its columns from the assistant general secretary of the Gas Workers' Union, that "We, as workers, have no objection whatever to a sound scheme of profit-sharing," and upon Mr. Livesey's admission before the Parliamentary enquiry, expresses the opinion that the exclusion of trades-unions is no essential part of this kind of copartnership.

To show that the hostility in this case between the company and the trades-union is an accident, and not necessary in a system of sharing profits and responsibilities, it quotes the case of another profit-sharing plan of Mr. Livesey himself, adopted at his instance by another London gas company. This is the Crystal Palace Gas Company, which is also under the management of Mr. George Livesey. It is governed by a sliding scale for the regulation of dividends and prices, like that of Mr. Livesey's other company, the South Metropolitan; and the directors have now voluntarily included labor in the benefits of the sliding scale, giving wages a bonus for every reduction on the price of gas. Almost the only difference between this plan and that of the South Metropolitan company is that no trades-union is in any way prohibited by it. *Labor Copartnership* also quotes an American instance, that of Mr. N. O. Nelson's co-operative enterprise at Leclaire, where, Mr. Nelson says, his "company encourages its men to belong to trades-unions, believing such organization has a useful influence, and that it is a necessary protection for wage-workers."

Under the bonus scheme of 1889, the annual bonus received by the workingmen in the South Metropolitan was five per cent a year for two years, the price of gas having been reduced from two shillings eightpence to two shillings threepence, or fifty-six cents per thousand, the price at which gas is now sold. Then good times came; coal rose and the price of gas rose, and the bonus was reduced to three per cent, with no complaint from the men. Then it rose again to four per cent a year, and five per cent; in 1895 it was six per cent, and in 1896 and 1897, each, seven and one-half per cent. At a meeting with the workingmen before the strike, when Mr. Livesey submitted to them the profit-sharing plan, one of the men said: "I have ever cast my longing eyes upon the (gas) stocks which realized such handsome sums, and should like to ask whether it is not possible to lower the quantity sold from £100 to £10 in order to give us a chance. There is nothing will bind us more to the concern than giving us a monetary interest in it." The same hope that facilities might be given them to become stockholders was expressed by another one of the men.

In pursuance of this hint, the directors of the company began a system of buying stock to be cut up into small lots and sold again to the employés, as well as the consumers; and in the four years following the adoption of the bonus scheme, the amount invested by employés, in £5 lots, was $21,405, over $5,000 a year. By judicious purchases by the directors in advance, stock was supplied to the workingmen at less than its market price at the time it was transferred to them. They thus got stock at 135 when it was quoted at 145, and the year before at 125 when quoted at 140.

In 1894 the company took another step forward. So far it had only shared in profits; but, as Mr. Livesey said to me, "profit-sharing is only a halfway house to copartnership." The directors in this year offered to increase the annual bonus to one and one-half per cent instead of one for each penny reduction in the price of gas, on condition that one-half of each man's annual bonus should be invested in the company's ordinary stock. Up to this time the men had been free to withdraw their bonus, but under the new plan only the half not invested in stock could be withdrawn, and a week's notice was required. The first year fifteen per cent of the men preferred the old scale, but gradually they all came in and accepted this arrangement. The bonus for 1897 was seven and one-half per cent on the wages of the men, and these averaged $7.50 a week. As the result of this bonus scheme of 1894, the investments by the men in the stock at the end of each year thereafter have been as follows:

1893	$ 21,405
1894	88,205
1895	128,210
1896	230,395
1897	345,000

The compulsory investment increased the voluntary investment. In three years, although the half bonus to be invested amounted to $105,000, the actual investment was $210,000, twice as much.

"The figures," says Mr. Livesey, "speak for themselves more eloquently than anything that can be said in their favor." There have been very few sales of their stock by the men. In a great many cases, husband and wife are joint holders. Sometimes a man would like to sell, but, as one of the directors said, "fortunately joined to the man is the woman, and if the man wanted to go on a spree, they required two signatures to the transfer. There the woman was the sheet-anchor." It is part of Mr. Livesey's policy to interest the women. "Take the opinion of your wives," he said to his men in one of his circulars about their profit sharing. Frequently when the company finds that the man wants to sell to get money for some good purpose, it advances him the cash and lets him keep his stock; and in every one of these cases the loan has been repaid.

Before the adoption of this measure of 1894, the men, though sharing in profits, became stockholders only as they chose to do so; but now all the employés of the company must capitalize one-half at least of their share of the profits. All employés must, therefore, be stockholders.

Mr. Livesey now thought the time was ripe for the greatest step of all in his program—admitting the workingmen to the board of direction. He describes his progress as having consisted of four steps:

First: The agreements for a fixed term of service.

Second: Profit-sharing, creating a union of interests.

Third: Shareholding or partnership in the profits and risks.

Fourth: "Workmen directors," or partnership in control.

Over this last feature of his program, Mr. Livesey had a hard fight with his stockholders.

There was not at last a note of dissent against profit-sharing, but "Workmen Directors" were too much. His own brother, an official of the company, at a meeting of the Institution of Gas Engineers, May 6, 1897, stated that with the first three of his brother's steps he heartily concurred, but not with the "Workman Director," and I found him still of the same mind. Mr. Livesey took the simple ground that the workingmen could not be admitted to the position of shareholders without sooner or later demanding the right to be represented on the board, and he thought it much better policy to give it than to have to give it. In 1893 the company needed to go to Parliament for some modifications of its charter, and Mr. Livesey tried to persuade its directors to allow him to get permission from Parliament for the election of workmen as directors, but they refused. Mr. Livesey, however, put up Mr. Mundella to move a "Workman Director" amendment; and this was favored by the Duke of Devonshire, Sir Michael Hicks Beach, Mr. Thomas Burt, one of the labor members, and Lord Morley, who was president of the co-operative congress at Plymouth in 1886. But the bill, for some reason or other, was withdrawn.

When later it became necessary to revive it,

Mr. Livesey said plainly, in substance, to his shareholders and directors: "If you want me to continue as your chairman, it must be upon my own terms. I cannot stay here unless the workingmen are allowed to have representatives of their own upon the board of direction." He was thus strenuous about the matter because of his conviction that only by such a measure of justice and consideration could the highest welfare of the company and its public be made secure. This time the shareholders and directors assented. As passed by Parliament, the act required that the company's workingmen should hold $200,000 of its stock before "Workmen Directors" could be elected; and a workman could not be eligible for the position of director unless he had $1,250 of stock. Mr. Livesey regarded this latter requirement as excessive, and in 1897 he succeeded in getting Parliament to pass an act reducing this amount to $500. "There are," Mr. Livesey told me, "a number of men owning more than this amount." The act stipulates that there shall be seven capitalist directors and three wage-earning directors. The workingmen altogether must hold $200,000 of the stock. When I was in London, the amount held by them was within $3,000 of this, in addition to $55,000 held by the salaried men. This year the necessary $200,000 has been already more than made up, and "Workmen Directors" will be elected.

One of the interesting features of this experiment is that profit-sharing has stimulated the men to increase their rate of saving out of their wages, so that their investments in stock and their deposits on interest have each been more than had been expected. In fact, the men have put so much of their money into the stock of the company, in addition to the compulsory investment of half their bonus, that they have reached the amount of investment required for the election of "Workmen Directors" two or three years sooner than was anticipated. All the men are not yet shareholders in their own names, as the amounts to their credit are yet, in some cases, too small even for a £5 purchase. These savings are accumulated in the hands of trustees and are credited to the purchase of stock. In two years more, says Mr. Livesey, every man in the company's employ will be a shareholder in his own name. At the end of 1897, 1,196 employés held in their own names $242,200 of stock, the market value of which was $351,190.

The "Workman Director" is an experiment, limited at first to three years; and it certainly will not be renewed at the expiration of that period unless the results justify it. Mr. Livesey's brother and other opponents among the directors insist that "the workmen are not yet educated up to such a position." Will the workingmen elect agitators or another kind of men to represent them? Mr. Livesey is firm in his faith in the men. He reports that they have been heard to say in their conversation with each other that they do not want on the board as their representative "any chattering workman." "We have had," he said in the *International Review* for March, 1896, "for the last six years a joint committee of workmen elected by ballot, and an equal number of officials nominated by the directors to manage the profit-sharing scheme, and they have, at the same time, dealt with any other questions relating to the workingmen that had arisen. So far, although the committee is large, and one-third retire every year, causing a good many changes, it has worked most satisfactorily without a hitch of any kind." This, he thinks, justifies him in his expectation that the workingmen will choose practical men as directors. One of the superintendents of the company at the meeting of the gas engineers, just spoken of, also reported that at one time he might have said that to have a committee of workmen sit down and discuss matters affecting the conduct of the undertaking would have been attended with grave difficulties; but it did work, and it worked admirably, and no one who had not had direct experience could understand how easily little difficulties were removed almost as soon as they appeared. At this meeting one of the company's officials mentioned a fact which bears on the competence of the men. "Ten years or so ago," he said, "the average stoker might be able to read, but it frequently happened that he could not write — I had very good evidence of this in obtaining signatures to certain proposals (some years ago) when I found that a large proportion of the men could not sign their names. Now it is rare to find a man who is not able to write, and the majority write well." With this extension of their education, he thought the men were putting to themselves the questions which, as a gas engineer, he knew gas engineers did: "Ought I not to have more interest in the prosperity of the undertaking for which I am working so hard?"

As the result of seven years' work, the amount of money which the employés of the company now have in its stock and on deposit with it was, in the middle of 1897, well over $400,000, about half in stock and half in savings. There then were 700 or 800 employé shareholders, and by the end of 1897 there were 1,196 of them. The synopsis of the company's report for the last half of 1897, in the *Co-operative News* of February 12, 1898, states that the employés hold stock of the value of $345,000, in addition to $185,000 on deposit at interest, making their accumulated savings then in the hands of the company, in round numbers, $530,000.

In his address before the gas engineers, Mr. Livesey told what the men thought of the bonus scheme as expressed in hundreds of letters from them which he had invited. Over and over again, these men say: "We wish it could be applied to other concerns; we wish other firms would adopt it." Mr. Livesey read the following letter:

"*Dear Sir:* I must say we" (that is the writer and his wife) "have derived great benefit from your profit-sharing system, and think it is one of the best methods ever started to bring masters and men together. My wife and I both agree about that. Eight years ago, we had not got a penny saved, but we have never drawn a farthing of our bonus, and have added considerably to it. Besides that, we leave two shillings a week for savings, and we are proud and happy to say that we have accumulated enough to buy a hundred-pound share" (worth one hundred and forty pounds) "after eight years. I am only a carman, and have a family to support. We only hope and trust it will continue, as we want to have something laid up for our old age. Having derived so much benefit ourselves, we would be extremely pleased to see it adopted in every firm throughout the country, and that is our true and honest opinion."

Mr. Livesey believes that this arrangement —crowning profit-sharing with the "Workman Director"—will mark a new departure in the relations of capital and labor. When every employé is an actual shareholder in his own name; when the men who have entrusted almost all their entire savings to the company are themselves trusted by the company with a share in its responsibilities, he believes that the beginning will have been made of the introduction of the new order in the industrial world.

As will be expected, the opposite view is held with intensity by many socialists and trades-unionists. The *Pall Mall Gazette* described the agreement or bond for a year's service as "bond slavery." In the letter already referred to, the secretary of the Gas Workers' Union says of the bonus scheme: "Any right thinking man will agree that it is nothing more or less than an intrigue to so bind the men as to leave them quite powerless so far as united action is concerned when any emergency may arise." Even the non-unionists who are pocketing the bonus cannot fail to note that these concessions were accelerated by the organization of the Gas Workers' Union. It is fair to the company to say that it was already proceeding in the direction of giving the men an interest in the company; and it is fair to the union to say that but for it the men would to-day probably still be without hundreds of thousands of dollars which they have won in addition to their standard wages. And one of the most distinguished of the younger professors of political economy in Great Britain said to me that he looked upon Mr. Livesey's experiment as very dangerous, as meant to "nobble" the workingmen, and give them a special interest apart from their class interest.

Mr. Livesey's opponents also point out that his tactics of interesting workingmen in the ownership of the gas works are erecting a formidable barricade in the path of municipal ownership. He is creating an army of small proprietors, who, as all experience has shown, will be most jealous and irritable enemies of any interference with their "vested rights." His critics, therefore, say that his experiment is directly contrary to the public interest, since he has smoothed the relations between employer and employés by helping to rivet on the shoulders of the people of London a burden which they ought not to be asked to bear—private monopoly—and has impeded their progress towards municipalization, which has done so well for Glasgow, Birmingham, Manchester, and other places. They also point out that Mr. Livesey has shown that in his opinion the bonus is not costing shareholders anything, for he argued in his speech before the London Chamber of Commerce, January 22, 1898, that "if workmen would enter into the spirit of the thing and do their best, they could easily save five per cent, or whatever the percentage might be, by lessening the cost of production." Mr. Livesey has himself made it an argument against municipal ownership of gas works that municipalities having no stock could not admit their employés, as his company has done, to participation in profits and management; and he has repeatedly avowed that he believes that the Employé Shareholder and Workman Director is an institution which will make socialism impossible, because it will give the wage-earner the possession of property.

But if this was a "capitalistic intrigue," it certainly did not have the support of the capitalists. It was, as we have shown, actually forced upon his directors and shareholders by Mr. Livesey, who is more an employé than a capitalist, as also his father was before him. "These gas works were the playground of my childhood," he said to me. The representative of another company, when Mr. Livesey described his scheme before the meeting of gas engineers, met it with the sarcasm that it seemed to him that "before long all the directors will be working in the retort-house, and the retort-house men will be sitting on the board." And at the meeting of the Gaslight & Coke Company of London, in February, 1890, the chairman stated that his company had not adopted the same scheme of bonuses as the South Metropolitan "because we thought a quasi-partnership with the men would involve the right to participate in the government of the company, and this was not to be thought of."

CHAPTER XI

THE WORK OF GENERATIONS

It is not a "mushroom growth," but a long, historic evolution, which has brought England to the point where it can offer the inspiring spectacles of labor copartnership. The oldest of the living co-operative societies—that of Govan Old Victualling in Scotland—was started in 1777. For nearly a hundred years the working people of Great Britain have been struggling towards this goal of self-employment and self-government in industry, to which Robert Owen opened the way in the early years of the century. As early as 1829 we read in the *Co-operator* of December 1st of that year: "There are about one hundred and thirty co-operative societies now established. Co-operative manufactures may be now purchased at the Co-operative Bazar, 19 Grenville street, Hatton Gardens." All of these failed. The causes of failure were, as explained by the editor of the *Co-operator*, bad business management; giving credit; intemperance; indifference of the working people, some members not dealing with their own stores; the interference of the chartists, who used the stores for propaganda.

Another fruitful failure in the early part of the century was that of the communities organized by Robert Owen and his followers. His social remedy was to establish co-operative communities of a more or less self-supporting character. These, as well those in America as in England, found economic conditions and the stage of character development of the people against them; and they all failed.

The third wave of co-operative production swept over England in the '40's. The Christian Socialists rode this wave. Profiting by the lesson of the Owenite failures in community life, they gave themselves to the establishment of "self-governing workshops." Their efforts are briefly described in an article in the *Economic Review* for July, 1894, on the copartnership of labor, by Henry Vivian and Aneurin Williams. In these shops the actual workers were to own or borrow all the capital, elect the committee of management from among themselves, and enjoy all the profits, subject only to two conditions—that every workshop must make a payment into the fund of the movement every year, and that at least one-third of the profit must be capitalized. These founders were well aware of the narrow field within which such societies could succeed, since the large masses of workers at the bottom, and those whose operations required great capital, could neither of them be helped in this way. But the Christian Socialists thought that there was field enough for them among the higher classes of workingmen, whose occupations required but small amounts of capital. They set up in business a number of societies of artisans of this kind.

The last number of the *Christian Socialist* published in London in 1851, contains a list of co-operative associations and stores. This shows fourteen workmen's associations in London, and one or more in twenty-one other places. There were flour-mill societies in fourteen places, and co-operative stores, one or more, in 146 communities.

The experiments in self-governing workshops, too, failed—all of them. But, as Mr. J. M. Ludlow has said, the failure was commercial, not moral. "No movement could be a moral failure which had connected with it such men as Frederic Denison Maurice, Charles Kingsley, Thomas Hughes, and Edward Vansittart Neale," and all the co-operators of to-day will add Mr. J. M. Ludlow himself.

Three mistakes had been made: First, the founders supplied practically all the money.

The men, therefore, had nothing of their own at stake. Second, no attempt was made to get picked men. These men, coming at haphazard, and with nothing of their own at risk, were, third, given full control over the workshops—that is, over other people's property. The inevitable result was quarrels and all kinds of narrowness and greed. Each of these three waves of failure brought its gift of needed lessons. As out of the failures of more than half a century of attempts at distributive co-operation, or stores, there finally came the Rochdale movement; so out of these failures of co-operative production has sprung labor copartnership, which has achieved the successes chronicled in these pages.

When a great thing is done, there are always to be found behind it great men. In co-operative production, as well as in co-operative distribution, there have been many great personalities, Kingsley, Maurice, Hughes, Holyoake; but one of them, Mr. E. Vansittart Neale, deserves special mention for the sacrifices he made of property and position, and the dauntless resoluteness with which he worked to wrest from calamitous failures the secret of the present success. Mr. Neale was a wealthy man who became interested in co-operative production through an advertisement which he saw of the Working Tailors' Association, one of those just referred to as having been organized by the Christian Socialists. He sought membership in the Society for Promoting Workingmen's Associations, and changed its policy, which had been cautious and tentative, into one much more aggressive. "The society had been doing nothing more ambitious," Thomas Hughes tells us, "than starting small productive associations of tailors, shoemakers and bakers, who could be mainly started, in the first instance, by the custom of themselves and their friends." Applications for advance of capital and for advice as to how to proceed, were pouring in upon them from numerous groups of artisans all over the country; but until Mr. Neale's arrival they had scarcely yielded to this pressure from the outside.

The new member, in sporting phrase, "forced the running." At considerable expense he started a co-operative store in London, and advanced the capital for two working builders' associations. Afterwards, at the time of the great lockout in 1851, in the engineering trade, which gave new interest to co-operative production as a social solution, Mr. Neale supplied the capital for putting a number of the locked out men at work in the Atlas Engineering Works, at Southwark. He confessed himself that he lost over $200,000 in these enterprises; and his total losses in co-operation are estimated by his friends at over $300,000. He had to give up his fine house at Mayfair and his chambers at Lincoln's Inn, taking a small house at Hampstead and economizing relentlessly. The associations which he had so chivalrously endowed failed. "Ninety-nine men out of every hundred," Judge Hughes says, "would have lost all enthusiasm, if not all belief, in co-operation under such circumstances, or, at any rate, would have become its 'candid friends.'" But with Mr. Neale this was but the beginning of the work, which he continued, and which was the foundation of the developments of labor copartnership which we are describing.

After the failure of the self-governing workshop experiment, Mr. Neale devoted himself almost entirely to the work of the Co-operative Union and its annual congresses, an organization within the co-operative movement for educational purposes. With many other thoughtful and advanced co-operators, he became dissatisfied with the almost complete absorption of the co-operative movement in the work of distribution. As the result of suggestions made by him, Mr. George Jacob Holyoake, Mr. E. O. Greening, and others, a special conference was held on the occasion of the Derby Congress, in 1884, and was attended by two hundred and fifty delegates and friends. This conference resulted in a resolution to form a propagandist committee and a propagandist union, to arouse workingmen and public opinion generally to the importance of the movement for making workers everywhere partners in their workshops, both as regards profits and shares. Mr. Neale was first president of the new organization, named the "Labor Association," with the sub-title "For promoting co-operative production based on the copartnership of the workers."

When Mr. Hughes received word of the proposed formation of the Labor Association, he wrote one of his friends: "I was glad to get yours with the news of the proposed union of productive societies for mutual aid—the best news I have had this many a long day."

The purpose of this association was to bring the co-operators back to the idea with which their movement was originally started. The program of the pioneers of Rochdale put forward as their ultimate object the formation of home colonies, in which, as Mr. Neale has said, "the workers, sustained by the fruits of their own labor, might be able, by wisely-organized institutions, to secure to themselves and their belongings, every advantage—educational, intellectual, moral, and social"—which could be attained by co-operation, aided by all the labor-saving marvels of modern industry. Mr. Neale quoted the results which in less than half a century grew out of their simple plan of emancipating themselves by buying groceries at wholesale to be sold to themselves at retail as an illustration to encourage us to believe in the practicableness and wisdom of their ultimate object—the complete self-employment and self-government of labor.

The Labor Association has remained what Mr. Neale described it to be at the beginning, essentially a propagandist body, which seeks only to form opinion, and thus to stimulate action. It has served, as he supposed it might, as a valuable center of union among the productive societies which its propaganda called forth; but it has not engaged in any productive enterprises of its own, nor supplied any money for others to do so. Mr. Neale outlined the mode of operations which the association would follow in this way:

"To form public opinion on the subject of associated labor by the following means, viz: The publication and supply of literature; the delivery of lectures, addresses, etc.; the holding of conferences of all classes of persons interested in the elevation of the worker.

"To assist workingmen to organize themselves for mutual self-employment.

"To enlist the active interest of the trade societies in the co-operative movement.

"To secure a united action of trades-unionists and co-operators for mutual benefit and progress.

"To give information generally on the position of co-operative workshops and the condition of the workers."

The form which has been taken by the self-employment movement under the auspices of this new effort differs from the experiments of Owen's day or those fostered by the Christian Socialists. The idea of a self-governing workshop, an independent, individualized group, self-owned, self-directed, and self-absorbed, has been as definitely abandoned as the earlier idea of a colony. The reaction of the pioneers against the demand of capital for all the profits had taken at first the shape of a counter-demand that labor should have all the profits; but now the movement has swung to an equilibrium between these two extremes, and stands on the proposition that labor, capital, and the public (the consumer) are all parties in interest, and are all to be given the right to share. Mr. E.

Vansittart Neale, for instance, at first insisted that all the profits should go to labor; but later he argued against his own earlier view, and advocated the importance of giving a share of the profits to the customer. The early plan of giving only a fixed return to capital and all the the rest to labor, together with the whole control, did not succeed. In striking contrast to this failure had been the successes of the consumers' organizations. But the Labor Association is as conspicuous in its success as its predecessor, the Christian Society for Promoting Workingmen's Associations, was in its failure. The time had become ripe. As put by Miss Sybella Gurney, in "Sixty Years of Co-operation:" "It required the growth of a better generation—a generation trained in the store, friendly society and trades-union movements, as well as at school, before a large measure of success could be achieved." Since its organization, the capital, business, and profits of labor copartnership have increased tenfold. Mr. Neale and his associates used to think, as he expressed it, that it was "a sin against the true spirit of co-operation for capital to ask for more than five per cent." But copartnership now gives capital not only its fixed dividend, but also a share in the profits. Thus we find the Kettering Boot and Shoe society, in the report for the last half of 1897, giving capital, in addition to its fixed rate of five per cent, a share in the profits amounting to two and one-half per cent more.

In the new movement the workers receive a share of the profit, not the whole; they enjoy a share of the control, not the whole. The societies have not been started for workmen by outsiders designing to do good, but by workmen for themselves, (*a*) sometimes with and sometimes without outside help; (*b*) in a few cases by employers, passing through profit-sharing to copartnership; and (*c*) by co-operative societies. Another fundamental difference is, that these societies are loss-sharing as well as profit-sharing. The workers, being part owners and part managers, have to be responsible as well for failures as for successes.

Subsidized co-operation has had its day. Many things have been learned from past experiments, and one of them is that gifts of money, or even the loan of all the capital required, is the surest of all ways to kill a co-operative society.

In commenting on a productive society which allowed itself to be absorbed by the Scottish Wholesale society, *Labor Copartnership* said: "This is another instance showing that where the impulse to co-operation comes from above and not from the workers themselves, successful copartnership can hardly be attained. This mill was started for, and not by, the workers. The result shows that they never realized their opportunity." The object of the Labor Association, like that of the Agricultural Organization Society started in Ireland by Mr. Plunkett, was to bring to the help of the working people the time, experience, and good will of those who desired to give them an opportunity to realize their aspirations for a better life. Those who had time gave it to those who had none. Those who had knowledge of the world and of the methods of organization gave it to those who were entirely inexperienced. Those who had enthusiasm gave it to those who had less and needed to be kindled. They cleared away legal obstacles which existed in England to organizations of workingmen. They obtained from Parliament legal protection for the property of co-operative societies and other improvements in the law. This and other work of general education and business training of workmen were the things that made the recent growth of co-operative production possible; and in this work Mr. Neale and his friends were pioneers and among the most active workers. They believed that if as much as this were done for the working people, they would do for themselves all that remained to be done. They were confident that as, by the Rochdale plan, capital was accumulated for the workmen-customers out of the surplus profits of their purchases, so by labor copartnership the workmen-producers could accumulate capital out of the surplus profits of their production. The result has justified them, as in Ireland the results of a similar policy are justifying Mr. Plunkett and his associates.

The character of the work done by the Labor Association would be entirely misunderstood if it were not emphasized that this society has no position of authority in the movement. It is purely a voluntary organization, which holds no papal relations of any kind to the copartnership societies of Great Britain. Not even a majority of them belong to it, though the best ones do. It has no power of discipline. Societies may join or not, remain or not, contribute to its funds or not, as they prefer. But it is a movement towards federation among the productive co-operative societies, and it has followed the same line of evolution that was taken in the distributive movement. First, there were the distributive stores; then came the English Co-operative Wholesale and the Scottish Co-operative Wholesale societies, and the Co-operative Union, which are federations of retail societies. The Labor Association and its running mate, the Productive Federation, to be described later, are, loosely, corresponding examples of federation among the productive societies. The productive movement under this initiative and inspiration has taken on a new life. The report of the Labor Association gives the following figures of the growth of the business carried on. The pounds sterling of the original table are changed into dollars at the rate of $5 to £1:

	1895	1896
Number of societies,	127	152
Sales for year,	$9,299,380	$10,824,010
Capital, share, loan and reserve,	4,576,510	5,388,580
Profits,	471,525	564,955
Losses,	11,480	60,350
Dividend on wages,	71,175	80,415

The increase for the year 1896 is, approximately, twenty per cent in the number of societies, sixteen per cent in the sales, eighteen per cent in capital, ten per cent in profits, after deducting the losses, and fourteen per cent in the share of profit paid as dividend to wages. These *Labor Copartnership* figures as an all-round increase of sixteen per cent. Last year there was an increase all round of thirty per cent, but in the year before ten per cent only.

The following shows the growth for thirteen years:

	1883	1893	1896
Societies,	15	108	152
Sales,	$803,755	$6,463,440	$10,824,010
Capital,	517,180	3,199,420	5,388,580
Profits,	45,155	338,315	564,955
Losses,	570	14,920	60,350
Dividend on wages,	Unknown	41,415	80,415

There is here an increase of about twelve-fold in the thirteen years. It is Ireland which gives most of the last year's increase. There are twenty-two more Irish societies than a year ago; in England, four more; in Scotland, one less. All three countries show increase in sales and capital. Profits show an increase in Ireland and Scotland, a decrease in England. *Labor Copartnership* describes the year as one of substantial, though not sensational, growth in England and Scotland, and characterizes Ireland's progress as "splendid."

Mr. R. A. Anderson, secretary of the Irish Agricultural Organization Society, sends me the following comparative statement showing the number of societies in existence in Ireland on March 31, 1897, and on March 18, 1898:

	1897	1898	Increase
1. Co-operative dairy societies,	84		
Auxiliaries,	10—94	131	37
2. Agricultural societies,	45	71	26
3. Co-operative banks,	3	13	10
4. Miscellaneous societies,	4	11	7
5. Federations for purchase and sale,	2	2	—
Total,	148	228	80

The "dividend on wages" seems small compared with the total profits. It is to be remembered that, besides this dividend on wages, the workingmen get the benefits of the provident, educational, and other funds. The committeemen, also, receive pay. In many cases the workers are all shareholders, and receive interest and dividends, and these are the fruits largely of dividends upon wages previously got. Then, too, as we have found in our progress through the copartnership world, wages are often higher, hours shorter, holidays more numerous, and employment steadier, than in the trade at large. *Labor Copartnership* estimates that the real money-profit to labor, including all the above considerations, is at least double the amount credited as dividend to wages.

The Labor Association has been pushing the copartnership movement by issuing a large number of pamphlets, by extensive agitation through public meetings over the whole of Great Britain, and by giving assistance and advice to workingmen who announce themselves desirous of forming productive societies. During the year over two hundred and fifty lectures are delivered by the Labor Association's representatives, chiefly in the great industrial centers of England and Scotland, and some 250,000 documents distributed. In addition to this, a number of important local exhibitions of co-operative products are organized, many conferences held with persons interested in the industrial problem and the establishment of new productive works. Its lecturers have been invited to Oxford and Cambridge to address the students there. A monthly *Labor Copartnership* is published, under the editorship of Miss Sybella Gurney, for whom the invitations of the life of ease that might be hers, have no charm in comparison with those of this cause—"a cause," one of her associates said, "which is worth giving one's life to, since it is educating the people for the coming change, whether it is to be peaceful or violent." This journal is a record of the advanced development of the co-operative movement in Great Britain, which no student of its progress can afford to be without. The *Irish Homestead*, of Dublin, does a similar work, with signal ability, for the co-operators of Ireland.

The voluntary relations existing between the Labor Association and the copartnership societies of the country are illustrated by the fact that although there are 152 societies which can be classified as copartnership, only fifty-four of them are affiliated with the association. Its officers are influential men. The president is

Mr. George Jacob Holyoake; its legal adviser is Mr. J. M. Ludlow, one of the Christian Socialist group of the 50's for many years registrar of friendly societies in the government service. Among its vice-presidents are the Bishop of Durham, Earl Grey, the Earl of Stamford, Professor Marshall, Hon. T. A. Brassey, and Tom Mann. The most active member of the executive force is Mr. Aneurin Williams, a man of means, who comes of a manufacturing family, who has devoted much of his time and money to this effort to place workingmen in better relations to their work, and start them on the road to becoming captains of industry. In an article on "The Copartnership of Labor in 1896," in the Co-operative Productive Federation Year Book of 1897, Mr. Williams asks and answers the question as to what is the ultimate goal of the movement:

"One naturally asks, What is the ultimate goal of such a movement? Does it offer a complete solution of the social question? There are, perhaps, few of its advocates who claim that it does, but certainly there can be no complete solution without it. There is a wide sphere of work which can be best done by the state, perhaps can only be done by the state. There is a similar sphere for the local governing bodies. But the whole extent of ordinary manufactures would seem to be outside what either the state or the municipality can do with advantage. We at least claim that they can best be carried on by voluntary co-operative associations, not indeed managed solely in the interests of their own employés, but in which those employés, with their expert knowledge and their direct interest in the result, shall be a powerful element. We look forward, therefore, to the multiplication of these copartnership societies more and more rapidly as time goes on. But not only so; we look forward to their more and more complete federation, so that competition among them, in any bad sense of the word, may be avoided, and common action be taken for all common purposes. In this way an organization of industry will be —or rather is being—built up, which shall combine the advantages of central action on all great common interests with the local self-government of each society of workers, and a responsibility coming closely home to every individual."

The figures which are given of the growth of the movement show a rate of progress as rapid as that of the distributive or store movement in its earlier years. Between 1883 and 1896 there has been an increase of over tenfold, in the number of the concerns, in the capital employed, in the sales and the net profits. In the last year, 1896, for which official figures have been published, exactly the same number of societies have been registered for production as for distribution. The opponents of copartnership are fond of saying it has "no future." A movement that has increased tenfold in thirteen years, and now has a capital of $5,000,000, an annual product of $10,000,000, and an annual profit of $500,000, only needs to carry such a present along with it; it can get along without a future.

A great deal is said about failures; but, one success proves more than any number of failures. The industries which are now represented by copartnership societies, cover so wide a range, including as they do, hosiery, boots, tinplate, woolen goods, cotton cloth, tweeds, engineering, leather, printing, watchmaking, cutlery, creameries, building, clothing, pottery, and many others, that it seems entirely reasonable to believe that the movement has in it the vigor for perpetuation and wide extension.

Ten years ago the copartnership societies were as one to a hundred compared to distributive societies, but now they are as one to ten. This result is the fruit of hard work and a great deal of devoted self-sacrifice. The members and officers of the Labor Association would be the last to limit their ideal to the "self-governing workshop." "We have no objection," the salutatory of *Labor Copartnership* said, "to part of the capital being owned by persons not employed and part of the control being exercised by them. * * * Modern industry more and more requires larger amounts of capital than the actual workers can themselves provide, at any rate at starting; and, further, the presence of outside members, whether individuals, stores, or labor organizations, should act as a moderating and impartial element." Copartnership says only that labor shall share in ownership, management, and results. It seeks to harmonize all five of the interests involved in production—the employé, the employer, the consumer, the trades-union, the general public. It asks for all workers such a voice in the management of their own industry as democracy demands that the people should always have in their own affairs. Composed of cooperators, and springing up within the cooperative movement, it seeks this, first, in co-operative territory, but it considers it to be its duty to promote this policy even in private business and in government employ.

CHAPTER XII

THE NEW JOY OF LABOR

London has been called a great co-operative desert. Many attempts have been made to establish co-operative stores; but, although the population and the market are the greatest in the country, these experiments have never been anything like the success which has been had elsewhere, as at Leeds, Manchester, or Glasgow. It is outside of London that one must look for the star examples of co-operative effort. The causes of this lie on the surface, and hardly need to be described. The population of London is fluctuating, the distractions are innumerable. The first condition of co-operative success, as the Rev. Dr. Maurice has pointed out, is mutual confidence; and that is a plant of slow growth, and one of almost impossible growth, where people have no opportunity to know each other. The Labor Association has had more success outside of London than in it.

Under its auspices some years ago, quite a

number of workshops were started in London, but the earlier ones were blighted with the exception of a bookbinders' society. This, after many vicissitudes, appears to have at last struck its roots strongly enough to live. It was founded in 1885, and from the first it has had a severe struggle, owing largely to the fact that the usual customers of bookbinders—retail booksellers and stationers—with a very few exceptions, refused to give work, no doubt on account of the society being co-operative. It has, therefore, had to build up its business on the orders of individual customers; and such orders are so small and so far apart, that even a small workshop needs a very large circle of supporters. Co-operative sympathizers, however, have been enlisting in larger and larger numbers. Several societies now send all their binding and repairing for libraries to be done here. The roll of patrons is a long one, and in it are some interesting names. The Amalgamated Society of Engineers, the Hon. T. A. Brassey, Toynbee Hall, Library of Balliol College, Sir Frederic Pollock, Rev. Dr. Martineau, Guy's Hospital, are among them. Some large publishers now send here their better class of work. The society makes a specialty of binding old and rare books. It pays trades-union wages, and has the eight-hours day, which it gave three months before it was asked for by the trade. Though it is a recognized usage in the trade to dock for holidays, the co-operative society pays for all holidays, including the jubilees. Its latest reports are of increasing prosperity. In the Labor Association there is a cocoa works at Thames-Ditton, in opposition to which the Co-operative Wholesale Society has established cocoa works of its own. There is also a leather society in London.

A most promising copartnership enterprise in London, under the auspices of the Labor Association, is a building society under the title of General Builders, Limited. Three years were taken in the work of organizing this society. It was not allowed to start until it had obtained $5,000 of capital. It has eight hundred members, and employs forty to fifty workmen, all of whom are members of the society, and sharers in profits or losses and in management. It has $10,000 of capital in shares and loans. There are one or two middle-class shareholders, but ninety-five per cent of the share capital is held by workingmen. The first suggestion of it came, as has so often been the case in the productive movement, from a strike. Mr. Henry Vivian, of the Labor Association, went among the London joiners when they were out of work, and proposed to them that they should organize a co-operative building society to employ themselves. A meeting was called, all the trades-unions in the building trades were notified to send representatives, and a general circular was issued and distributed throughout the trades. Since the start was made the society has had a promising career. It operates in all branches of the building trade—brick, stone, concrete, terra-cotta, iron, plumbing, gas-fitting, painting, and all kinds of woodwork. Its hours are fifty a week during the summer, and it was able to keep busy all last winter. Trades-union wages are paid, and sometimes a little more.

I saw a block of ten houses which it had built in The Avenue at Tottenham, and also a fine bakery at Woodgreen, which it was putting up for the co-operative bakery there. The general secretary, and manager, of the society pointed out in the work, with a great deal of pride, the evidences of the superiority of their methods to those of the "jerry" builder. "Our men are not of the kind 'jerry' builders would employ," he said. As a practical detail, where the ordinary builder would lay ten courses of brick he pointed out that the co-operators laid eleven in the same space of wall, giving their patrons brick instead of

mortar. Enough has been made to pay five per cent on loans and on the shares, and a reserve is to be accumulated against future losses and to guarantee the payment of interest on capital before the members pay any bonus to themselves. The future prospects of the society are good. A number of architects sympathize with the effort because they belong to that enlightened element in the profession which has wanted for three hundred years to get into direct touch with the workers in order to be able to restore something of the independence and art feeling which distinguish the architectural work of the artisans during the days of the guilds. The society has had contracts given to it without competition because the architects found that it worked up to the standard. In one case, an architect told them they were doing the job too well; there could be no profit to them in it. There are large capitalists definitely known to be favorable to such an extent that future work is assured. This society now has seventeen branches in different parts of London. They have been successful enough to feel justified in building themselves an establishment for a central joinery works, which will include an educational room for lectures; and the whole, with the land, will cost $10,000 to $15,000. In the balance-sheet for the first half of 1897 the society state that they could use $10,000 more capital than they have.

To supply such needs for capital as this, a society called the Co-operative Productive Federation has been utilized as a sort of "running mate" for the Labor Association. The work of the latter is propagandist; that of the Federation is financial. The Co-operative Productive Federation is a society of societies. Its members are those of the productive copartnership societies which desire to assist, by federation, in developing the movement by advancing capital to other societies which need it and can use it to advantage. Its secretary is Mr. T. Blandford, a devoted worker, writer and speaker for copartnership. Fifty copartnership societies are now members, which have a capital of about three-quarters of a million dollars. The Federation has a capital contributed by them of $5,000. Every productive society, on joining the Federation, makes itself responsible for one $5 share for every five of its members, and also a small proportion of its profits. The work of the Federation is to receive money on loan from those societies which have a surplus, and to lend it to those societies which are in need of capital. The Federation pays four per cent, and charges five per cent. The margin of one per cent is being accumulated as a reserve, or, as it is called, an investment insurance fund, to make good any losses. The expenses of the Federation are very slight and are paid by a levy on the societies composing it, and not taken from its margin of one per cent between the interest paid and interest received. The loans have been so judiciously made, and the solvency of the co-operative movement averages so high, that there have been so far no losses to trench on this fund. The larger societies which have no need of borrowing are joining for the purpose of being able to help the younger societies. The membership includes some of the oldest, most important, and best known productive associations.

Up to December, 1897, $20,000 had been deposited with the society, and invested by it, and it was calling for larger deposits. The society has proved an excellent medium for both co-operative societies and individuals to invest loan capital, and also to those who, though wishful to help, have no time to make the necessary inquiry. By making their loans through the Federation, they save all this trouble and have their risk reduced to the minimum. Although so largely animated by this helpful spirit, the Federation is a strictly business concern. It asks for money only on the basis of paying interest, and lends it only where there is no reason to doubt its prompt repayment. The co-operators do not believe much in philanthropy. As their paper, *Labor Copartnership*, has said: "Nine-tenths of philanthropy and one-tenth of self-help will never make a co-operative workshop a success; but if the proportion is reversed, satisfactory results may be expected." A plan is being urged by Mr. Henry W. Wolff for the establishment of a bank in connection with the Productive Federation, on the model of the People's Banks on the Continent, which he has so interestingly described in his book; and there is little doubt that such an institution would be as useful an adjunct as the co-operative movement of Great Britain could have.

One of the best, and at the same time, most interesting and characteristic things done by the Labor Association and the Productive Federation, which work continually together, is the exhibition in the yearly co-operative festival. There have been ten annual national co-operative festivals, in which they helped, at the Crystal Palace. They began twelve years ago as a little flower show, in South Kensington, attended by two thousand people. Up to four years ago, it was a one-day affair only. Now it fills almost a week, and is an important public event. The *Times*, *Standard*, the

Chronicle, *Spectator*, and other leading London papers, give it column after column of notice through all the five days of its session. The exhibition is of the products of copartnership works, and those of the capitalistic factories of the English Co-operative Wholesale Society are, therefore, withheld. The exhibit is not one of the great commercial advertising affairs which are called "expositions," but a display of goods made in places where industry is conducted under humane conditions, and where men co-operate instead of individually competing for places, wages, and profits. Exhibitions of flowers, fruit, and vegetables are part of these festivals, and gardening all over Great Britain has received a lasting stimulus from this show, which is the largest in the world. There are athletics and open-air sports and amusements for the thousands of children who attend. The greatest audiences of the co-operative world are gathered to hear the best co-operative speakers. Co-operative music and literature are developed by the words and music which are written every year for these festivals. The singing is done by co-operative choirs. In 1888 there were 4,300 singers, and in 1896, 9,521. The attendance has grown from 27,169 in 1888 until in 1896 it reached 41,500. In 1897 it was 34,695.

We are all familiar with the outcries raised by Ruskin and Carlyle, and the other apostles of a better life for the people, with regard to the distressing conditions of what Matthew Arnold calls "the lower class brutalized." These festivals are an illustration of the way in which the co-operative movement has sought to do its share of the reform called for. Mr. E. O. Greening, who was their originator, has, for the past ten years, given up taking his regular holiday in order to organize these festivals for working co-operators. July and August, which he used to give to seaside journeys, he spends in London, busily engaged in committee meetings, council meetings, organizing, canvassing, working harder during this holiday season than in any other part of the year. The reasons which lead him to attach so much importance to these festivals and flower shows and musical celebrations, have been given by Mr. Greening in *Labor Copartnership*. They illustrate admirably the co-operative spirit:

"I walk down dreary London streets of workmen's houses. Every house is like every other one in its unredeemed ugliness. Yet each one has a small forecourt in front, or yard behind, which, my little knowledge of gardening teaches me, could be filled with a wealth of floral beauty. A trail of ivy taken out of a hedge, or a rootlet of Virginia creeper properly planted and watered and tended, would cover these homes with verdant loveliness. A few seeds or cuttings would fill front or back window, porch, and yard walls with glory. How can I wait for government to pass a law prescribing what each man shall grow, and appointing inspectors at good salaries to chivy the people by fines into compliance? How can I, on the other hand, leave the people to their own individualistic ineptitude? I must have an industrial flower show, great enough to arouse attention and wonder and enthusiasm. But it must be a show which demonstrates the power of working people to endow themselves with gardens of taste.

"In like manner, I am struck by the absence of music from the daily lives of our people. The open doors and windows send out no sweet notes of song, no concert of voices uniting in harmony, to delight us as we pass. The father and mother think they have no time to practice or to help their children to learn.

"Yet my limited knowledge of music tells me how easy, how inexpensive, and how expeditious is the acquirement of tonic-sol-fa or ordinary methods of music and singing, sufficient to enable the young folk who live in all these houses to fill them with delightful melody.

"And when I consider the power of subscribed pence under co-operative combination, I know and feel that only the will and agreement are needed to enable our industrial co-operators to transform their own lives and the life of the community from dreariness to fullness of delight. Our annual festival is an object-lesson to teach this truth by a great and striking illustration. But if the festival is to perform its perfect work, the men and women who have worked with me in creating it, must now begin to contemplate a new and greater development. Every co-operative store, every co-operative workshop must have its recreative and social departments; must provide for the leisure hours of labor the means of delightful recreation as varied as the tastes of men and women; must be missionaries of education in the means of happiness; in a word, must be centers of everything pleasant and profitable in the highest sense of the words. We have to teach our people and to persuade our societies to carry out this program. Who will help?"

The festival is something done; and in an address before the last co-operative festival, the tenth, Mr. J. M. Ludlow, survivor of the gallant band of Christian Socialists, put before the co-operators the idea of something new to be done—the creation of a great permanent playground. "I have seen," he said, "with great pleasure the movement for co-operative seaside homes, but these, however valuable, could not take the place of what I suggest. I know that many of the larger societies have their own cricket matches and athletic sports, and have acquired fields for the purpose; that others have spacious halls for concerts and other recreative purposes. But what I should wish to see, would be a sort of peaceful Bisley, a place of cricket fields and tennis courts, football fields and golf links, with water at hand for some boating in summer and much skating in winter, a place where the jaded co-operator may go for a few days' play—honest play, God-fearing play, for we all know that play itself may be dishonest—and return to his work refreshed, a new man. Such a place—though it should certainly include at least a good concert and assembly room for evening use—would not, I consider, of necessity interfere in the least with this festival—might, on the contrary, be made to work in with it and help it. I throw out this idea for your consideration, knowing perfectly well that I shall never see it worked out (indeed, I am afraid we, most of us, know that of all eggs in the world, co-operative eggs, although we trust they are the best of any, are perhaps those which take the longest hatching). But it seems to me that within the next thirty years, or even twenty, the thing might be done. I can even seem to see in the distance a great co-operative park extending over half a county, in which co-operators might spend, say, a month's holiday with their families in houses or hotels belonging to the great central co-operative authority of the day. For I am well convinced that, great as are the strides which co-operation has taken in various directions in my lifetime, it is but, as it were, Tom Thumb at the very outset of his career."

Another plan which is still in the air is worth mention for the light it throws on the co-operative spirit. An organization is being mooted for co-operative visits and holidays among the members of the co-operative societies. This proposal originated among the members of the Women's Co-operative Guild, which has ten thousand members, organized into 204 branches, with branch meetings, district and sectional conferences, and with a three days' annual gathering, with the object of studying the principles of the movement and how to put them into practice. This women's guild is teaching its members when they go to the stores to ask with regard to the commodities offered them: "Is this co-operative made?" in order to spur the stores into doing their duty in extending productive co-operation. Among the members of this guild, the suggestion is being discussed whether some simple kind of organization shall be formed to enable friends who live in the country or at the seaside to receive visits of members or members' children from the cities, and to organize the whole business of reciprocal hospitality among the co-operators. In pushing the plan, one of its advocates proposes that a "Holiday and Visiting" column be opened in the co-operative newspapers, and gives these specimens of the kind of items that might be put into it:

CLASS I.

Mrs. A., living in the country near Bradford, would take two children for holidays. Has one girl of ten.

Mrs. B. and baby would be glad to have a fortnight's change in the country or by sea.

Mrs. D. offers accommodation to any young co-operator visiting London.

CLASS II.

Mrs. A., Blackpool, offers a month at the sea for a month at Leeds.

Mr. and Mrs. B. wish to spend a month in Belgium with a co-operative family. Hospitality in return.

Mrs. C. wishes to send her daughter of eighteen, who works in a mill, for a fortnight's change. Mill household preferred.

CLASS III.

A party is arranging to visit Familistere at Guise Probable cost, $25. Four vacancies.

A small party of co-operators in the North are anxious

to visit Shieldhall, and propose going by boat to Glasgow. Will Scotch friends communicate? Return hospitality to Manchester and Liverpool.

A party of French co-operators would be glad of hospitality.

"Where the co-operative movement shows itself most is in the children," Mr. James Deans says. "They are better fed, better housed, better behaved, and brighter. The Co-operative movement shows more in them than it does in the parents." These co-operative festivals, flower shows, parks of "half a county," and reciprocal hospitalities promise a still better childhood, and fatherhood, and motherhood for the working people of Great Britain. "Co-operation," George Jacob Holyoake has said, "is business saddled with morality." It is something even finer. Co-operation is business democratizing itself, garlanded, dancing, and set to music, the Ten Commandments and the Golden Rule.

CHAPTER XIII

WORKMAN VERSUS BUSINESS MAN

The "business man" has been trying conclusions in several severe struggles with the co-operators, and the co-operators get the best of it. They show courage, finesse and resources, at least equal to those of the competitive captains of industry. The fiercest of these battles, so far, is the "boycott" against the co-operators, now on in Scotland, with the result that Scotch co-operative progress has been going on by leaps and bounds since it began. This boycott has been very bitter, and has gone so far as to war on the sons and daughters of co-operators, who have been told by private employers that they cannot retain their positions unless all the members of their family discontinue connection with co-operative societies. The boycotters tried to prevent the co-operative societies, many of which do their own butchering, from buying cattle at the cattle auctions, and induced the auctioneers to refuse to take bids of co-operative buyers. But the co-operators went outside the market to the farmers, and succeeded in getting all the cattle they wanted. The only result has been that a great many more Scotch societies are butchering their own meat than did so previously.

Not one of the co-operative butchers' stores was closed for an hour. The co-operators of Scotland met this attack with something more than mere commercial inspiration. "Are the descendants of the Covenanters and Reformers to run away from a regiment of grocers?" I heard one of their orators, Mr. James Deans, exclaim, to the thunderous applause of a co-operative audience. The net outcome of the boycott has been to give the co-operative movement a great advertisement, out of which it is reaping a very important gain in its membership and business.

The co-operators know so well the benefits of agitation of any kind that they seek them on every opportunity, and sometimes in amusing ways. One of them described to me a boycott in the early days in Manchester against the Salford store. An important provision merchant turned against the society when it began to buy its supplies directly from the producers instead of patronizing him, as it had been doing. He began a vigorous attack in the newspapers, using the information which he had gained through many years' connection with the society, as ammunition for his articles. My informant was appointed to answer these attacks. He was specially instructed to take the greatest pains to get the worst of the argument, so as to keep the other man going. This he did, although he said it was with no little difficulty that he was able to hold his good arguments out of sight. The effect of the controversy was that whereas new members had been coming in at the rate of ten a week, they began to come in at the rate of two hundred, and the society gained three thousand during the debate. The butter and bacon merchant was enjoying his polemic triumph hugely when some one told him of the ruse that was being played, and his letters stopped abruptly.

The co-operators can put on a braver front if need be. The railroads in England, as in America, are among the real rulers of the country, controlling, as they do so largely, economic destinies, and with them the co-operators have more than once come to an issue. "Whatever grievances we may have to complain of," Mr. Plunkett said in discussing this question before the Irish Co-operative Conference of last November, "they are incidental to a system in which unorganized individuals do business with powerfully organized corporations." A story told me in Manchester shows how much advantage the co-operators gain by being able to pit organization against organization.

Some years ago, one of the great railroads in England announced to its employés that they must choose between their membership in co-operative societies and their employment by the railroad. The station-masters gave the notice along the line. This was shrewdly designed as a death-blow to the co-operative movement in that part of the country, and must have been suggested to the railroad by some of its shippers, who were losing their business on account of the growth of the co-operative stores. In one town alone there were in the seven thousand members of the co-operative store six thousand who were railroad men. If this order was enforced and the men, to save their positions on the railroad, had to give up their membership, the co-operative store would practically cease to exist. In some of the towns along the road, entire committees of management would be swept away.

There is an organization which has for its function, among other duties, to deal with such attacks as this—the Co-operative Union, a body under whose auspices the co-operative congresses are held. Its able secretary is Mr. J.

C. Gray, who got his initiation in co-operative work as an official of the Hebden Bridge fustian cutters' society. The Union is the only co-operative organization covering the entire kingdom. Its membership includes upward of one thousand societies, representing a million and a quarter members, or six-sevenths of the co-operative population. Its work is educational, parliamentary, and propagandist. It gives legal, commercial, and other advice to its members as needed. In one of its pamphlets prepared by Mr. John Allan and Mr. J. C. Gray, its secretary, it is said:

"The traders of the country, jealous of our success, have frequently made strenuous efforts to induce public bodies, railway companies, and large employers of labor, to compel their employés either to give up their connection with co-operation or otherwise lose their employment. In every case where the services of the Union have been called in, we are glad to say they have been instrumental in checkmating this atrocious attempt on the part of selfish competitors to coerce the workingman in regard to the manner in which he shall occupy the hours remaining to him apart from his labor."

This railroad attack upon the co-operators in its employ was one of the emergencies for which the Co-operative Union exists, and its officials promptly answered the challenge. One of them sought an interview with the general manager of the railroad. He obtained it with difficulty. When he announced his errand, the manager pooh-poohed it.

"I represent," the co-operative official said, "in this matter all the co-operative societies of Great Britain, doing a business of $250,000,000 a year, with a million members."

Still the manager was contemptuous. "We must dictate," he said, "our own policy to our own employés."

"Yes," replied the co-operator, "'and our members must and will dictate their own policy to themselves."

The manager did not see how this interested him as an outsider. The co-operator moved another pawn:

"We can at once divert $500,000 worth of business from your lines; and we will do as much more next week, and every week, until you leave our men free to decide for themselves what they will do with their own time and in their own private affairs."

The manager's attitude changed. "I see," he said, "that there is something in this matter which I had not understood." There was indeed.

The result of the interview was that the

offensive order was withdrawn, and no attempt has ever been made to re-introduce it. One of the co-operative societies involved held several hundred shares of the stock of this railroad. Had the railroad persisted in its course, the society would have transferred this stock in single share lots to an equal number of workingmen, and sent them all in a body to attend the next meeting of the railroad stockholders, where, if they had the average temper of the British citizen, they would have been pretty certain to make themselves heard.

Upon another occasion, a signalman of one of the principal railroads was discharged because he was secretary of a co-operative society. The same Co-operative Union official took charge of the case. He saw the man. "Can you vouch," he said, "that you have never used any of the railroad's time for your co-operative work?" The man said that he could. "Then," said the co-operator, "we will fight your case through to the end." They did, and the man was reinstated.

The ability of the co-operative societies to divert traffic was here again an effectual remedy, which only had to be threatened. I asked the gentleman who managed these difficult negotiations for the co-operators what he would have done if there had not been, as there might not have been in America, this competition among the railroads of which he could take advantage. He replied that one remedy which he might have used would be to bring the men out on a strike. They were so large a proportion of the total number of the railroad employés, and the sympathy between the trades-unionists and the co-operators was so strong, that the strike would certainly have become general. The purchase of railroad stock to qualify co-operators to attend the meetings of stockholders would have also been a part of their tactics. A movement which is embarrassed, as the co-operative movement is, by a glut of capital, could have made such a demonstration to any extent necessary. It is not a hazardous guess that if all other remedies failed, the co-operators would deal with the railroads by political means. The railroads can easily crowd the co-operative societies into a state of mind which would throw the overwhelming majority of the voters among their million and a half of members on the side of nationalization.

One thing, it was admitted to me by a leader in the war of the private traders, must be conceded to the co-operators. They have taught the people to buy for cash, to save their money, and to study investments. He told the story— a very old one among co-operators, but new to me—of a woman who was being condoled with by her parson upon the death of her husband. Her husband had been "buried and hearsed" by the co-operative society of which he had been a member, one of the newest ventures of which was the business of undertaking. To the condolences of the minister his widow replied: "I have a mecht o' consolation. I will get a dividend out o' him yet"—from her share of the profits in the funeral.

It is the fashion among outside critics of the co-operative movement to depreciate the business ability of workingmen, and on this foundation are built the willing prophecies that co-operation must fail, but the question of the ability of workingmen to organize successful business seems to be settling itself. Lord Winchilsea's statement before the co-operative congress, already alluded to, is pertinent here; and the copartnership people quote with great effect a very remarkable declaration made at a meeting of the Amalgamated Society of Engineers by Sir Benjamin C. Brown, the eminent engineer employer, who expressed the opinion "that the workmen should really become the owners of the works in which they were employed, and when they became owners, they would be on the board of directors. There was no reason why, of one thousand men, they should not have some who were qualified to manage large works. It could not be done at once; it would require a number of years, but what was there to hinder them more than any one else from learning it? If this were done, he did not see any reason why, in the future, a large number of workers should not become owners, and the limited companies should not be in the hands of the working classes. There was a large number of houses in that district which were owned by workmen, and something like the same capital would be required. About $750 per man was the capital necessary to start an engineering business." The co-operative movement can easily supply $750 per man, and many times that, for any large works it desires to organize co-operatively. It can do so, as Sir Benjamin says, quite as easily as it can furnish the same or a larger amount to workingmen to become the owners of their own homes, as has been done in thousands of cases.

Mrs. Sidney Webb in her book gives an eloquent answer to the argument that the small salaries which are paid managers in the co-operative stores are inadequate to attract the necessary brains and knowledge, or, as she wittily says, "that calibre of brain which secures sound financial operations on the part of

Messrs. Baring Brothers, or the intimate knowledge of the cotton market that has distinguished Mr. Steelstrand,"—the cotton broker who failed for millions in trying to run a great cotton "corner." Mrs. Webb continues: "The rising tide of co-operative enterprise, the steady progress of the store system, with its federal institutions, adding year by year to the multiplicity of their operations in industry and commerce, are a sufficient answer to this biased or theoretical view of human motive. The good will of a great community, the political power and social influence equitably earned by the able and energetic official of a powerful and growing organization, have proved as efficient a form of remuneration as the unknown gains and lawless expenditure of the capitalist *entrepreneur*, or the exorbitant salaries given by middle-class shareholders to middle-class officials, consequent on an extravagant estimate of the conventional outlay due to social position." I have not met in the world of private enterprise a finer set of young business managers than the intelligent, alert, industrious, and devoted young men of the labor copartnership enterprises of Great Britain.

The facts of co-operation have put an end forever to the superstition that great commercial, financial, and administrative ability was the monopoly of a few or a class. The flax-spinners of Leeds, in fifty years, beginning with a few shillings, have built up a business which now has two millions and a half capital, makes sales of $5,000,000 a year, with a profit of $750,000. These mill workers successfully initiated a flour mill, and then went on with the establishment of boot factories, stores, abattoirs, building societies, etc., until the result is one of the most successful and most varied businesses of modern times—the most successful local co-operative store in Great Britain. "Mechanics," says Mr. George Jacob Holyoake, in his Jubilee History of Leeds, "with all tradition, all experience, all the prophets of social and commercial disaster, against them, built up the largest co-operative society that has ever existed."

The Frame-Makers and Gilders' Association is one of the workingmen's associations cited by Mr. Benjamin Jones to prove his contention that such efforts must be unsuccessful. He gives several pages of his book "Co-operative Production," to this instance; but Mr. J. M. Ludlow shows that so long as the work of the society was carried on by workingmen for workingmen, it was extraordinarily successful. Its downfall began when it admitted a business man and made him master director. One of the most profitable and flourishing enterprises in the co-operative productive movement is the Co-operative Printing Society of Manchester and London. It was started by some trades-unionists who were encouraged to undertake it by the spectacle they saw in Manchester of three printing shops which had been successfully run by a timber merchant, a coal dealer, and a cotton broker. The result has proved these workingmen printers had as much ability as these outsiders to manage a printing business.

Plain, ordinary wage-earners have proved that they can save, accumulate, lend and borrow, organize, finance, anticipate markets and wrest great businesses from the hands of those who had the advantage of position, capital, inherited aptitudes. Business ability is no monopoly of the middle class, nor of exceptional individuals in any class. Most millionaires boast that they sprang from the working people. The success of co-operation proves that they left behind them as much ability in the multitude as theirs, and that this ability can be organized as successfully through democratic methods of association as under individualistic capitalism, or even more successfully. Capitalism has proved that industrial genius was to be found in individuals of all classes.

The joint stock movement and co-operation have proved that this genius can be operated through bodies of men. The next and last step in this progress is to prove that the largest body of all—the people—can administer it.

Weavers, tailors, mill-workers, shoemakers, cannot be considered destitute of business ability when, entering upon the task which Mrs. Sidney Webb epigrammatically calls "emancipation by penny subscription," they have, in the thirty-four years, between 1862 and 1895, accumulated a capital of $106,650,000, doing a business of $275,500,000 in 1895, and millions more now, with profits in 1895 of $26,000,000 and increasing every year since, and a membership of 1,430,340. The total sales of these establishments between 1862 and 1895, inclusive, were $4,078,801,705, and their profits in that period have been $360,377,840. This they have done in the face of every obstacle from the outside that could be thrown in their way by prejudice, wealth, rank, conservatism, and selfish interest. But, despite all these, they have succeeded. Mr. T. P. Gill, member of Parliament, writes in the New York *World*, November 28, 1897:

"Now all this immense and varied scheme of commercial enterprise has been carried on from first to last by workingmen. This is undoubtedly the most remarkable thing about it. Workingmen are among the presidents and managers of the wholesale societies; workingmen have organized, systematized, classified, and directed the various departments of this business, and these leaders of the concern have been selected and appointed by a constituency of workingmen according to the principles of the most complete democracy."

And this is but the beginning. The co-operative stores and factories have still but a fraction of the business even of the co-operators. The Co-operative Union publishes a little tract showing that they could, by carrying co-operation forward into the supply of their dwellings —displacing the landlord—and into commerce, workshops, farms, mines, and factories, increase their income to $7,500,000,000, or three times that which they now receive. This business success of the co-operative workingmen enlarges the conception of democracy. It shows that industrial democracy can be a fact whenever the people will it. The desire for property is a universal attribute of mankind, and aptitudes to manage property, like "honor and fame, from no condition rise." In fact, this is the threshold truth to a greater truth which democracy still has to disclose; which is, that property, business, and capital will never be properly managed until the entire people have a share in ownership, management, and results. The administration of the material side of the human and divine life demands, not less surely than the political and religious sides, the utterance of all voices and all energies.

Certainly co-operative enterprise, with all its failures, has no reason to shrink from comparison with private businesses, in which ninety out of every one hundred ventures fail. Mr. George Thomson, of Huddersfield, who has converted his workingmen into the owners of the establishment which once belonged to him, told me that in eleven years he had not had one cent of bad debts from co-operative societies. He believes that co-operative business is going to be the fittest that will survive in the competitions of the business world, because its work is better and more economically done; because co-operative workmen are steadier; because the absence of adulteration, misrepresentation, and other trade tricks makes it more acceptable to the consumer, and its copartnership makes it better for the workmen. It is able to get its pick of labor. It is not quite true, as one enthusiastic co-operator said to me, that co-operation has no strikes, no losses, no riots, no panics, but it is much more nearly true than it would be of private industry.

The one point in which private industry seems to have an overwhelming advantage in the competition with the co-operators to-day is its vast superiority in capital. But on every hand one finds in Co-operation land, that capital is rolling in upon co-operators as rapidly as they can use it, or even faster. Heckmondwike store, for instance, at a recent members' meeting, had pressed upon it the emergency of what to do with its surplus capital. It had on hand —and it is not by any means one of the largest stores in the movement—$300,000 of money to find investment for. It seems heretical to suggest doubt of the economic superiority of large capital in these days of gigantic consolidations; but it is still to be proved that capital cannot be too large for economic and successful administration. The consolidation of the flourmills at Minneapolis in the hands of one syndicate, although they have remained under exactly the same management as before, has resulted in the discontinuance of dividends.

The workingmen in some ways can stand the pressure of bad times and competition better than capitalists, who have apparently many times their financial strength. Capitalists are more cowardly, for one thing, than the workingmen. The co-operators of Kettering, I was told, could stand two years without profit. So might capitalists. The difference between the capitalist and the workingman would sometimes be that the capitalist would not stand it while the workingman would. The capitalist motto is to run profits and cut losses. I asked Mr. J. M. Ludlow what reason he had for thinking that co-operation would survive in a business world like ours, where merciless competition was the rule, and where the resources of cooperation, great as they were, seemed so insignificant in comparison with the total against them. He replied that the co-operators would survive because they had "the most grit." The workingman has another advantage. Where he has a share in ownership and control, he can work his capital himself. This the capitalist cannot do. His capital has to be employed for him by others. The workman capitalist cannot only be content with less of profit than the investing capitalist, but he can get through with no profit at all so long as his work brings him enough to keep him alive. The workman capitalist, in other words, has a hand as well as capital; the mere capitalist has no hand.

The publicity of co-operation must be counted one of its buttresses. This gives it the basis for a confidence as broad as the constituency, with a further advantage of suggestions of policy, etc., from all the points of view of all its members. An incident illustrating this democratic confidence of co-operators in their institutions, was told me by Mr. George Thomson. A rumor started in the local press there that the manager of the Huddersfield co-operative store had decamped with $75,000. This store, and all co-operative stores, even the great English Wholesale, have a large proportion of their resources in withdrawable capital—in some cases ninety per cent. But this disturbing rumor produced not a ripple of a run. Co-operators save enormously in not being obliged to advertise as private concerns do. While they have no privileged market among the co-operators, and while they command, at best, but a small part of the possible market among the co-operators, they still have therein one that is sure and constantly growing. I learned of no case where they advertised for outside trade. The co-operators sell for cash and buy for cash, or on time so short as to be equivalent to cash. The directors of the great Leeds store, at the very beginning, when they were poor and weak, went into the market and paid more for grain than the other mills. It was policy to buy not in the cheapest, but the best market. Commenting on this, Mr. George Jacob Holyoake says, in his "Jubilee History" of the Leeds store: "The common practice of commercial business is to buy in the cheapest market and sell in the dearest. The Leeds directors followed the better rule of buying in the best market, and supplying the poorest member with the best quality of food, which the working people in Leeds had never had before." When their members, craving the adulterated white flour on which they had been reared, demanded to be supplied with it, the directors had the courage and the far-sighted sagacity to refuse, even at the momentary cost of losing custom.

There is no sentiment in business, we are told continually by the average business man; but in co-operative business there is a great deal of sentiment, and this has been in the past, and will be in the future, one of the greatest springs of strength and success. The sentiment of co-operative loyalty, demanding co-operative products, has been already shown, in the war waged by the private traders of Scotland, to be a tower of strength to the movement; and this loyalty is as yet, even in the co-operative movement, in its infancy. One of the plans of the Co-operative Union for meeting a private war, if that should be begun, in England as in Scotland, is to appeal to the conscience of co-operative buyers, and organize a co-operative market which instead of the goods of private industry, "sweated" and hostile to co-operation shall prefer co-operative goods made under proper conditions, and strengthening the co-operative movement by their sale. Co-operators are susceptible to such appeals and would respond in large numbers, even if it involved sacrifice temporarily; but no such motive can ever come into view for the support of private industry. It is there absolutely non-existent.

"Co-operation will succeed economically," Mr. George Jacob Holyoake said to me, "because, among other reasons, it is supported by enthusiasm." "The co-operative workingman," says Mr. E. O. Greening, "knows that he has a better chance for expansion of life, and he will put his heart into his work in a way inconceivable by the ordinary worker." Mr. Greening adds, as another element in the forces making for the economic survival of co-operation, the vast store of outside capital waiting for investment. He stated that there are hundreds of co-operative societies which can get any amount of capital they might desire against any coalition of private manufacturers. These concerns are now refusing money, cutting down the rate of interest they pay their own investors; but if they chose to borrow it or needed capital, they could get all they wanted at a low rate of interest.

It was not the private traders fighting him, but the professors, the literary reformers, and the "men of the world," who assured me that the workingman had no ability for business. The private traders have the evidence of his business ability in the bruises they have suffered at his hands. They content themselves with explaining it away as merely temporary, "due," as one of them said to me, "to the leadership of fanatics willing to do any amount of work for almost no pay. When this generation of cheap fanatics disappears, co-operation will disappear." But I found that these "cheap fanatics" had gathered about them a staff of willing successors ready to take up their work, satisfied that if the pay was a little low there was a higher compensation in helping to make the world better and in getting the sympathetic approval of their fellow men.

CHAPTER XIV

TWO KINDS OF "STORE DEMOCRACY"

There are two great central "store democracies" in Great Britain, the English Co-operative Wholesale Society and the Scottish Co-operative Wholesale Society, which buy and manufacture goods to be sold to the retail co-operative stores if these choose to buy. The opponents of labor copartnership give the system they advocate—that of the English Wholesale—the title of "store democracy," or "consumer's democracy," or "Federalist school." Their spokesman is Mr. Benjamin Jones, the very capable manager of the London branch of the English Co-operative Wholesale Society, whose book, "Co-operative Production," was written to extol its method, and depreciates copartnership production. Mrs. Sidney Webb, in her book on the "Co-operative Movement in Great Britain," defends this "store" theory, or, as she calls it, the "democratic" or "federalized" system.

But there is also a "store democracy," a

"Federalist school," which favors labor copartnership — the Scotch. The English Wholesale Society is moving away from labor copartnership; the Scotch toward it. Both of these are "federal" bodies; each is a society of societies. In the English there are 1,046 societies which are shareholders; in the Scotch, 283. Each of these Wholesales is an independent society in itself, having its own capital, officers, rules, management and business. The Wholesales have no monopoly of the co-operative market. No co-operative society need buy a shilling's worth of them unless it wants to.

No one of the constituent societies has any authority over its Wholesale except as explained below, nor has it toward any of them, theoretically. But their magnitude and wealth, and the superior initiative of their directors, who are continually engaged in large co-operative transactions, while the directors of the constituent societies are often men who have their own living to make outside, give the Wholesales, as a matter of fact, a predominance.

The co-operative stores of each district hold meetings periodically to decide questions of business and policy. In these district meetings the Wholesale directors are represented by two of their own number; and with their wider experience and central prestige, they find it an easy matter usually to control the local delegates. Nominally, the Wholesale is under the control of the delegates chosen by the societies which hold shares in it, and for whose convenience it was constituted; but practically, I was assured by its critics, popular control is gradually becoming a mere name. The central government has become so large that its own public cannot deal with it. Another grip which the Wholesale has on its constituent societies is that these have invested in it or deposited with it large sums of money; and they are, therefore, and inevitably, extremely susceptible to its suggestions and its will.

The Scottish Wholesale is far more significant as a co-operative illustration than the English. The Scotch has a business of $21,-000,000 in a population of only four millions of people, while the English has one of $59,-000,000 in a population of thirty millions—relatively much less. The Scotch society was increasing at the end of 1897 at the rate of 14.8 per cent, the English at the rate of only 7⅜ per cent. Owing to the more general education of the people and their greater thrift and intelligence, co-operation has, from the beginning, flourished more in Scotland than in England, and in the last year or two, the growth of Scotch co-operation has been almost embarrass-ingly rapid, due in part to the war which is being made upon it by the private traders. Another reason given for the gain in Scotland is, that the first generation brought up under the new education law has recently been coming into active life.

The English Wholesale Society at Manchester favored profit-sharing at the beginning, as in its Leicester boot and shoe works; but when the society grew to be a millionaire, there came the millionaire spirit, and profit-sharing died out. Rochdale had the same experience. There was profit-sharing there at the beginning; and its success, Mr. Holyoake tells us, was "brilliant while it lasted, and would have continued so but for the cupidity of outside shareholders, who put profit-sharing to death. In their demented impetuosity to become rich, they sacrificed the chance of honor and honorable profit." This was in the spinning-mill at Mitchell Hey. The official who is now in charge of the great bank of the Co-operative Wholesale Society at Manchester, and who was one of the early stockholders in the Mitchell Hey mill, described to me the early enthusiasm of himself and his associates for profit-sharing. But he said their success brought in a rush of new stockholders, like the recruits who are now embarrassing the co-operative movement all over Scotland, and profit-sharing was swamped. "It could not be proposed now," he said; "the board do not believe in it." But he thought there was sure to be a new wave of this enthusiasm, and profit-sharing would come again. Its effect upon the working people, he believes, will be the same that the French Revolution had upon the peasantry of France; in giving them property it will make them thrifty and steady.

The English Co-operative Wholesale Society engages very extensively in productive co-operation, but it has eliminated from this every trace of profit-sharing or copartnership, except so far as the better wages and conditions which it aims to give are in effect a modified form of profit-sharing. This society is a continuation, after a long break, of a premature wholesale society which was founded by the Christian Socialists, and in which Mr. E. Vansittart Neale made his heaviest pecuniary sacrifices to the co-operative cause. No. 55 of the *Christian Socialist*, in 1851, contains the "Address of the Central Co-operative Agency," established "as a legal and financial institution for aiding the formation of stores and associations for buying and selling on their behalf, and ultimately for organizing credit and interchange between them." Of this central co-operative agency,

Mr. Neale and Mr. Hughes were trustees. The variance between the spirit in which this prototype was founded and that which now animates the English Wholesale in its repudiation of its early ideal of copartnership, may be seen in the description given by Mr. J. M. Ludlow in his "Notes of a Co-operative Tour," in No. 48 of the *Christian Socialist*, of the Salford co-operative store, run by "a little band of heroes." "I do not suppose," he says, "there is a co-operative association in England more deserving of support than this one. They are men who place moral considerations above all others. Beyond wages they divide nothing, but lay out all in setting men to work." There were, in that day, a number of such societies formed to give the people employment, beginning with the cheapest trades. They were called "Redemption Societies."

The policy of the English Wholesale is avowedly competitive, not only in its relations toward its workingmen, but toward other co-operative societies. We have several times noticed its invasion of the field of other co-operators, as in the Irish creamery business. It is pursuing an acknowledged policy of either swallowing any successful co-operative manufactory at once, or starting competition with it. When the Wholesale was begun it was an-

nounced that it would not interfere with any
other co-operative productive society already at
work. But this has been changed. In his
book on co-operative production, Mr. Jones
says (p. 760) that it is " a settled principle that
whenever the co-operators enrolled in the retail
societies, and through them in the wholesale
societies, desire to produce the articles they require for their own consumption, they will do
so not only when the field is unoccupied,
but even if there is a co-operative productive
society already in the field; provided this
society is not of the federal type of association." As a matter of fact, the English
Wholesale does not refrain from competition
even with associations of "the federal type."
The Irish dairies, as well as the cocoa works
and the boot and shoe works, and other enterprises of the copartnership world, are of the
federal type; but the Wholesale has entered
into vigorous competition with all of them. It
has gone so far as to refuse to admit to membership a co-operative society engaged in copartnership production, on the ground that its
goods compete with the manufactures of the
Wholesale; and yet this society had engaged
in this industry before the Wholesale, and was
selling two-thirds of its product through the
Wholesale.

An unfriendly critic, Mr. Joseph Ackland, in an article on "The Failure of Co-operation," in the *Economic Review* of July, 1897, says that it is "quite evident, therefore, that the English Wholesale Society is working on the principle of centralization. It aims at bringing the whole co-operative movement of England and Wales (there is a similar society for Scotland) within its own organization." He points out also that this "vast central organization, supported by the 'tied' stores, is not the limit of the ambition of these daring innovators. A large export trade is to be cultivated." He points out this will bring them into competition with co-operators in the countries to which they export.

The feeling of the English Wholesale is so strong against labor copartnership that it will not take part in the exhibitions of the Co-operative Festivals because the Labor Association shows its productions there. This is matched by the displeasure with which some co-operators in the Paris International Co-operative Congress viewed the huge pyramid of soap which the English Wholesale had there. "What are they doing here?" was their audible question.

When at Manchester, I visited the soap works which the English Wholesale has established at

Irlam. The buildings cover nearly three acres. Fifty different kinds of soap were made, and the sales were $2,500 a day in October, although the works were only started in March. The mechanical arrangements of the establishment leave nothing to be desired. The main floor has been made the height of the floor of the English freight cars, and by a switch the cars can be brought to the doors to receive shipments. The rosin which is used comes in the Wholesale's own ships, direct from Savannah, Georgia, to the dock at the Irlam works, which are on the Manchester ship canal. "There is n't in Europe its equal," the superintendent assured me. He said that the wages paid in the establishment would average from fifteen to twenty per cent more than those paid by their immediate competitors, but there was no other approach to profit-sharing.

The Annual of the Co-operative Wholesale Societies for 1898 shows that the English Wholesale is now engaged in the manufacture of biscuits, cocoa, butter, preserves, sweets, boots and shoes, soap, candles, woolen goods, ready-made and "bespoke" clothing, flour, lard, furniture, shirts, mantles, and underclothing, and has a printing department. It has in these works 5,214 employés, out of a total of 8,647 in its employ.

The wonderful success of the store movement has pulled co-operation into channels too exclusively commercial, and has obscured the productive philosophy which was really the mainspring of the movement at its inception. The stores founded by the Rochdale pioneers, and, before them, by the still earlier workers in the co-operative field, were not for commercial purposes, but as a means of raising the condition of the workingman. "To assist other societies" is one of the phrases in the program set forth by the Rochdale pioneers in 1844. The end has been almost eclipsed by the brilliant success of the means. The copartnership school, as pointed out by Mr. R. Halstead, in a tract published by the Labor Association, on "Co-operative Production," tries not only to frame a new economic system, but to make a new economic man; and it looks with alarm upon the Wholesale system of co-operative production as, in his words, "a huge working-class capitalism."

"Tom" Hughes was unsparing in his denunciation of the English Wholesale for its attitude toward labor copartnership. He could not forgive it for investing its surplus capital in consols, Manchester canal bonds, immense buildings, and outside enterprises, and refusing the help of its money to productive societies attempting to carry out the principles of the founders of the movement. I saw a number of his letters on this subject, and was given permission to quote them. In one he declared the managers of the English Wholesale were "tradesmen at heart, and cannot rise to the feeling that men should be put before profits. This is the danger to our movement, and I am anxious about it." In another he says: "I hold, with you, that unless through a well-worked system of higher education, our members can be made to understand and live up to their principles, we shall drift sadly to leeward." And in still another he wrote: "I look upon the apostasy of the Wholesale and the consequent demoralization of many of our large societies as a much more serious danger to co-operation than you and other loyal co-operators seem to regard it; and it would be impossible for me to attend the congress and not to denounce it." "Please God," he said again, "it (the co-operative movement) will be the salvation of England's industrial life; but not in my time, and only after bitter sufferings for the backsliders of this generation."

The opponents of labor copartnership who inaccurately and, therefore, "unscientifically" call their school of productive co-operation the "store democracy," or "consumer's democracy," insist that the workingmen in its factories can best share in the results of their work by becoming members of "the store." One class this argument has evidently not convinced—these workers themselves. Very few of the workers in the English Wholesale's factories are members of the co-operative store, not ten per cent, was one estimate given to me. But almost all the copartnership factory workers in Kettering are members of the co-operative store of that town.

One very practical aspect of labor copartnership is the value to the workingmen of the discipline they get from it. Professor Marshall, who is a member of the Labor Association and sat on the Royal Commission of Labor, brought out this point in his examination of the representative of the Labor Association. He showed that the technical education of the workingman was much less important than that which came from being allowed to share in the management; and that copartnership production is a means of working towards a state of society in which the division between the upper and the lower classes will be removed.

If the copartnership works "sweat" their employés and violate trades-union conditions, and play the part of tyrannical masters, as is charged, why is it, the copartnership advocates ask, that the movement of the workingmen is to them and away from the works of the Wholesale, instead of in the opposite direction? Only one man has gone back from the copartnership shoe works in Leicester to the Wholesale works, while hundreds have gone in the other direction. "This argument," Mr. G. J. Holyoake said, "is like the argument of your pro-slavery people in America in their controversy with the abolitionists, that the slaves were better off under slavery, and did not desire their freedom. But they were never able to answer the question why it was that all the fugitive slaves ran from the South to the North, and why no black man ever ran from North to South."

The labor policy of the English Wholesale has never been more sharply challenged than in its own city of Manchester by the Bishop of Manchester, Doctor Moorhouse. At a co-operative meeting in January, 1898, in Manchester, the bishop said: "It sounds very well to say that in Great Britain and Ireland we have two hundred co-operative productive associations; that they employ no fewer than twenty-four thousand workmen, and that they are doing a trade of twenty millions a year; but when we come to inquire into the character of these associations, I fear we suffer an abatement of satisfaction, for the great mass of them are only associations of persons to provide capital, and they employ workers who do not share in that capital, nor in the profit that results from it. .... Remember, production by the co-operation of distributors is one thing" (referring to the English Co-operative Wholesale Society) "and production by the co-operation of producers is a different thing.

"I have carefully read from beginning to end all the reports of your sectional committees, and what have I found there? Why, over and over again that promising associations are languishing for want of capital. And I find also that you have in the distributive societies so much surplus capital that you are looking out for profitable investment. 'Profitable investment!' What do you mean by that? Do you mean investments that will bring you in a little money, or do you mean investments that will encourage the movement, that will lift the members of your class—the working class—to a higher status in intelligence, character, and prosperity? What do you mean? If you mean the former, you are dealing with your money like hucksters. And if you mean the latter, then I say you will not hesitate to invest your surplus capital in sound and well-considered associations for co-operative production * * * in which the workmen shall become profit-sharers—and ultimately part owners. * * * The ball set a-rolling will go on without stopping, until the co-operative means of carrying on great concerns has become the dominant system of this land."

How much need there is for these exhortations of the Bishop of Manchester is apparent from the fact stated by the Leicester builders, in their report to *Labor Copartnership* in February, 1898: "We have plenty of work at present. Our only drawback is want of capital. We had secured the contract to build a new free library for the city of Leicester, but have had to give it up as we had not the means to carry it through." Meanwhile, within a stone's throw almost of their headquarters are co-operative stores at their wits' end to know what to do with their surplus capital.

Earl Grey spoke on this subject with a thoroughness equal to that of the Bishop of Manchester, at a conference to discuss the international co-operative alliance, at London, in January, 1898. Earl Grey said that copartnership was "the most interesting and most useful phase of the co-operative movement. * * * It was an ennobling principle. * * * It developed that feeling of brotherhood that was so dear to the Christian Socialists. As civilization advanced, the world would be more inclined to accept the principle. Just as slavery gave way to serfdom, and serfdom to hiredom, so would hiredom give way to copartnership. Earlier experiments had failed because the people were not educated up to the principle."

"As you go north," an old co-operator in England said to me, "you will find co-operation more flourishing—and more niggardly." But the Scottish Co-operative Wholesale Society shares profits and management with its men and women. It is singular that in England, where it is discouraged by the most important co-operative organization, the Wholesale, the development of labor copartnership is very rapid, new copartnership works springing up on all sides; while in Scotland, where it is encouraged by the overshadowing central body, the Wholesale, there is almost no labor copartnership, except that of this body—and that is imperfect.

The rapid increase of its capital in Scotland, as elsewhere, is driving the co-operative movement into production. Part of the money of the Scottish Co-operative Wholesale has been expended—and no doubt wisely—in housing itself in a veritable palace of industry, in which it had been settled only a few months at the time of my visit. Its new quarters cover nearly an acre of ground, purchased at the rate of $100,000 an acre. The Wholesale is the largest commercial establishment in Scotland, and its premises are the finest commercial buildings in the country, and scarcely second in Glasgow to the best of its public edifices. The main front is 215 feet long; the height is five stories, besides a basement; and there is a tower 150 feet in height. The architecture is of the French classical renaissance of Louis the Fourteenth. The fine vestibule is approached through an ornamental wrought-iron gate; the floor is laid in mosaics, and the walls are lined and richly paneled with Sicilian marble. All through the co-operative world the superiority of the stores and factories over those of private industry is marked.

I went from this palace into the counting-room of one of the private traders who was leading the war against co-operation, and his dingy offices in contrast with the bright, airy, and beautiful premises of the Wholesale were almost like a tomb. Everywhere the British industrial democracy has brightened and lightened and beautified the habitat of work and business. When the democracy opens a new store or new factory, it makes a celebration of it, and goes dancing and singing through the doors of the new opportunity. The Wholesale's palace was opened with a great co-operative procession and music, with wagons bearing machinery in motion, groups of workers, displays of products, and a great dinner at which eight hundred people were entertained. These palaces stand where many remember the co-operative society to have made its beginning in a dark, little, upstairs shop. This achievement is all the work of poor workingmen, who have had success so rapid that the present cashier of the establishment, which is now doing a business of $21,000,000 a year, still a middle-aged man, began his career as the only clerk the society had.

The total number of the employés of the Scottish Wholesale is 4,605, and of these 3,746 are in its productive departments. All employés, whether in distributive or productive departments, are sharers in profits. The Wholesale was formed in 1868, and began the payment of bonus to its employés in 1870. The total amount paid as bonus up to the end of 1897 was $220,058, and of this $29,456 is on deposit with the Wholesale at three per cent interest. During the year 1896, the bonus paid the 2,569 workers in the factories was $12,980, which was nine per cent of the total profit made. Its system is not a full copartnership, but its movement has been steadily in that direction.

The president, Mr. Maxwell, before the Parliamentary Labor Commission, expressed his disbelief in the idea that giving the workers a share in the government of a concern would lead to want of discipline, and gave it as his conviction that co-operation will never be complete until employés have a voice as well as a share in profits. At the meetings of the society he has advocated that besides making all its employés sharers in the profits, it should also arrange that all of them should invest these profits in the stock of the society, so as to become members and entitled to be represented upon the board of managers—the system known as "compulsory capitalization" in the copartnership vocabulary. He was not able to carry his proposals, but the employés all have the privilege of so investing their share of the profits if they desire, and those who do become members are entitled to representation upon the board in a mode which will be explained. One of the particulars in which the Wholesale's system falls short of full copartnership is that its rules do not allow its employés to become members of the board of management. But as it gives the workingman a share in the profits and a voice in the management, it is properly ranked as one of the copartnership societies.

Its employés have not taken advantage, to any great extent, of the privilege of becoming members, although continuous effort is made to influence them in that direction. A short time ago the society issued a circular addressed to them, expressing its "regret that so few of the workers have taken shares," and asking them "to consider the advisability of becoming shareholders." The offer of the society is characterized as "a message to the workers," and it is deplored that it "has not been much more widely taken advantage of, especially when it is conceded on all hands that industrial partnership is one of the best, if not the best, means of interesting the workers in the management of business affairs and in the work in which they are personally engaged."

Attention is called to the fact that if the two thousand and more employés who are over twenty-one years of age, and eligible as shareholders, took advantage of their opportunity, they would be entitled, by the rules, to send fifteen delegates to the quarterly meetings. The employé shareholders are entitled to one representative, however few they may be, and also one for each one hundred and fifty who are shareholders. The Wholesale is a society of societies, and its own employés are the only individuals who are allowed to hold shares in it. This representation, it is pointed out, if occasion arose, would be able to plead the claims and protect the interests of the employés, and assist the delegates in dealing with matters affecting the workers. As it is, the employé shareholders number only 271, holding 4,263 shares out of 200,021, and have only two representatives. The circular gives full details of the conditions on which shares may be taken by the employés, has appended a form of application for shares, and closes with directions how they may, if they desire, obtain further information.

Some of the private manufacturers and business men of England are imitating this policy in dealing with employés. Among these, known as leaders in profit-sharing the world over, are William Thomson & Sons, of Huddersfield. The head of this concern, which is now the property of his former employés, Mr. George Thomson, has been the life-long friend and disciple of Ruskin, and was made by him the trustee of St. George's Guild. Veterans like Mr. George Jacob Holyoake anticipate that it will become general among private business men thus to admit their employés to a voice in the industry to which they yield their lives, and from which their living comes.

Labor copartnership makes an imposing array of authorities on its side. It can point to Robert Owen, who recognized the participation of labor in profit as a fundamental point in social reform and put it into practice at New Lanark in 1816. It was adopted by the immortal pioneers of Rochdale at the very beginning. They took the principle, says Mr. Holyoake, "into their democratic self-helping hands, and set up both store and workshop—purchasers and workers alike being sharers in the profit made by sale or labor."

Mr. Abraham Greenwood, the most distinguished living colleague of the Rochdale pioneers, said in an address at Leeds: "The association of labor, the division of the fruits of labor, or rather of the productions, between the producers in proportion to the amount and value of the work done by each, is the germ of an entire social transformation, which, by emancipating men from the servitude of wages, will gradually further and increase produce, and improve the economical position of the country."

The same leader in the congress of 1881 stated the fundamental principle of co-operation so finely that it must be given here to illuminate his advocacy of copartnership. It was "to teach mankind that, as humanity is one sole body, all we, being members of that body, are bound to labor for its development and seek to render its life more harmonious, vigorous, and active. .... Ask yourself as to every act you perform within the circle of family or country, If what I now do were done by and for all men, would it be beneficial or injurious to humanity? And if your conscience tells you it would be injurious, desist—desist even though it seems that an immediate advantage to your country or family would be the result."

"The best co-operators," says Mr. Holyoake again, "have never ceased to be of the same way of thinking. Every congress, when the question has been put, has voted for it. The chief statesmen of the day, peers, prelates, and the most distinguished of the clergy and nonconformist ministers, have declared in favor of equity to industry. Eminent political economists, as John Stuart Mill, Professors Fawcett, Marshall, Thorold Rogers, and Dr. Hodgson, have written on the same side." The great Schulze-Delitzsch is also one of the names to conjure with, with his oft-repeated words to the effect that "the only fully co-operative form of association is that constituted to carry on production, and that co-operative supply and co-operative credit can never aspire to be more than mere auxiliaries; and that productive co-operation must remain the one great ultimate aim of all genuine co-operators, as alone realizing the highest objects with which these co-operators set out upon their work.''

All the great names of the church are not on the side of Archbishop Magee, who was willing to go on record as saying that society would be disrupted by any attempt to apply the principles of Christianity in business. Dr. Westcott, Bishop of Durham, in his address to a co-operative demonstration at Tynemouth in 1890, said: "But however highly we rate these results (of the store), we must confess that in themselves they do not touch the real problem which lies before co-operators—the problem of the age, the problem of capital and labor; and therefore, gentlemen, though you may multiply your gains of this kind a hundredfold, though you may reach the utmost possible limit of cheapness and of purity for the benefit of the consumers, you will have to acknowledge that a great hope has been defeated, and a great work has been abandoned. You will have confined yourselves to commerce and exchange, and have left untouched the weightier and more difficult matters of industry and production. You will have ministered abundantly to individual interests, but you will not have effectually quickened the spirit of social service.

"I venture then to ask you to turn again to the program of the Rochdale pioneers, the heroic founders of living co-operation. The reform of distribution was for them the first step; and it was, in fact, the only possible step towards the reform of production."

There is in the whole muster-roll of co-operators, from the days of Robert Owen down to the present, not one great name which is not outspoken for the principle of copartnership. All that there is of prestige and renown, on the other side, is a few men who have made administrative successes in the management of the stores. But all the men who have commanded the attention of the world and whose names have become household words—Owen, Kingsley, "Tom" Hughes, Maurice, Holyoake—have never ceased to lend all the weight of their enthusiasm, and their genius, and their literary and moral prestige to this cause. Whenever the opponents of labor copartnership call the roll in speech, or essay, or editorial, of the great men who decorate the past and present of the movement, they invoke their own judges. It is their fate to have no other names to glory in than those who cry shame on them.

England has more than one "Grand Old

Man." Mr. George Jacob Holyoake and Mr. J. M. Ludlow, octogenarians, but both keeping unstained their youthful ideals and undimmed their early fire, unquenched by the waves of commercialism, which are drowning the enthusiasm of others; loyal to the original faith of the founders; always standing forth, in convention or in the press, as champions of the right of the workingmen—one president one year, the other the next, of the society which devotes itself to the propaganda of copartnership—are a spectacle to challenge the love and admiration of all, and are both grand old men.

CHAPTER XV

THE SCOTCH THISTLE

In giving a share of profits and management to its working people, the Scottish Wholesale is simply carrying further the policy of profit-sharing with which it began in its commercial department. The bonus is given to the workingmen and other employés at the same rate as to purchasing members. But only the profits of the manufacturing departments are divided among the workingmen in them. The employés of the distributive and productive establishments do not share in each other's gains; but all the profits of the various manufacturing enterprises are pooled and divided among all their workingmen. Co-operative enterprise does not do as private enterprise did at Pullman—causing the terrible strike there—compel the employés of one department to lose when that department happens to be unprofitable.

Only half of their bonus is paid to the workingmen. The other half is accumulated in what is called the "Bonus Loan Fund," and three per cent interest is allowed. It can be withdrawn only when the employé leaves the service of the company. There is not one woman, I was told, among the employés who has become a stockholder, though, of course, like all employés, the women all share in the profits. One explanation given me for this, by the manager of the shirt department, was that the girls did not accumulate their bonus because they all expected to be married. The managers are much disappointed that only 271 out of 4,605 have become stockholders. One of them commented upon what he called the inconsistency of the socialistic element among their workingmen, which is large. "They are all demanding that the workingmen shall become the owners of the means of production, but here, where they have the opportunity, they take no advantage of it."

One of the most impressive sights in the co-operative world is the great industrial colony of Shieldhall, outside of Glasgow, in which the Scottish Co-operative Wholesale Society is carrying on its productive enterprises. After the usual difficulties and failures at the beginning, the experiment of the society in manufacturing on its own account in Glasgow was so successful that the members made up their minds to organize production upon a great scale. Accumulating capital pushed them into this investment; they were impelled also by the certainty of profits, of getting superior commodities, and of striking a good blow for the elevation of the working people.

The site they chose is a very attractive one, on the banks of the Clyde, the same river where Owen, ninety years earlier, made his memorable experiment. The directors bought twelve acres of land, at a cost of $2,500 an acre. This was only ten years ago, but already the ground is nearly covered with factories turning out goods to the amount of over $5,000,000 a year. The buildings are arranged in three parallel lines running north and south. There are flowers and grass in front of the factories, and trim roads around them; an ornamental gateway and a clock tower give a touch of decoration to the architecture otherwise simple and businesslike. All the arrangements for light and air, sanitation and power, are of the most modern description.

Sixteen distinct trades are grouped here, each organized as a separate business, with hours, wages, and conditions conforming to the requirements of the union of its trade, and all pooling profits for the benefit of all the workers in them. "We do not think we are overstating the case," declares the "Hand Book" of the society for 1898, "when we say that nowhere else in this country, or in any other, are so many different kinds of industrial operations carried on practically within the same gate." The boot and shoe factory here is the largest in Scotland. The hours are fifty-three a week; the highest trades-union wages are paid; and overtime is paid at the rate of time and a quarter, which is not yet given by the trade generally. One of the features of the administration is the extended use of labor-saving appliances.

It must be counted a proof of business ability that the workingmen running the Scottish Wholesale, as one of their first enterprises, succeeded in the manufacture of shirts and ready-made garments, though these are among the most keenly cut trades in Glasgow, as everywhere. They had to compete against the "sweater," who had on his side the longest hours and the lowest wages. Mr. James Deans, speaking of this at the co-operative exhibition at the Crystal Palace in 1895, said: "It was the most terribly 'sweated' industry, and it was the first the Scottish Wholesale touched. They had given better wages, beautiful workshops, short hours, and withal they found they could turn out shirts and compete in price with the 'sweater.' Such was the power of justice to the worker."

The clothing business at Shieldhall is carried on in a building two hundred feet long by seventy wide, with a floor space of fourteen thousand square feet. The manager of the shirt department, Mrs. Kelley, is a woman who began eighteen years ago for the Scottish Wholesale, with three girls and herself. This was the first productive department started. There are now 209 employés, all girls, and almost all children of co-operators. In these bright and pleasant rooms no one sits in "unwomanly rags, with fingers weary and worn, with eyelids heavy and red." This, like the clothing manufacture, was a very unpromising industry at the start from the money-making point of view; but it has proved one of the most successful ventures. It began with an investment of $415 and a weekly wage-bill of $34. It now has an invested capital of $22,880 and a wage-bill of more than $19,000 a year. The account of this department, which is given by the society itself, confesses that it was influenced more or less in this effort by a philanthropic desire to improve the labor conditions of a very depressed class. There was sentiment in the motives which led it to undertake the manufacture of shirts, and there came with the sentiment financial success.

A few months after the start, the president

of the society was able to say at a business meeting of stockholders: "The fact that we are in a position to pay fair wages to our shirtmakers, and make a fair profit from this class of work, clearly proves to me at least that Tom Hood's dolorous 'Song of the Shirt' has no meaning to our seamstresses. To you, the shareholders, it must be a source of gratification that while you are making a profit you are also making a large number of your fellow beings happier and more contented than they have hitherto been." "The bright and airy rooms of this factory," says the society's account, "with its long rows of busy, cheerful workers, are surely the antithesis of Hood's pathetic picture." This department pays one shilling ninepence a dozen for piece work, for which the "sweaters" outside pay one shilling threepence; and two shillings threepence a dozen for piece work for which the outsiders pay only one shilling ninepence. At Shieldhall the operators are given the power and have machines that run fifteen hundred stitches a minute. . The "sweaters" charge their workers for steam, and oil, and thread.

Many of the co-operative shirtmaking girls live at home, but those who have to go to lodgings get their bed for two shillings sixpence a week. They can have their dinner in the co-operative restaurant for eightpence, which gives them soup, meat and potatoes, and can eat for one shilling sixpence a day. I was told that if I went to the private shirt factories in Glasgow I would not be allowed admittance, but I did not put this to the proof. These houses do a good deal of outside work, and their employés are running around with large bundles on their backs, while the co-operative seamstresses have no home work, no carrying of bundles, and have all the advantages of a fine, well-lighted and perfectly provided factory, with meals at cost at their own restaurant, recreation room, and other advantages.

The majority of the societies in Scotland now send their printing and bookbinding to the printing department at Shieldhall, which in ten years has had to have its accommodations doubled four times. The highest trades-union wages are paid, and the men have never known a day's idleness for want of employment. The printers have never struck, but there has been a lithographers' strike in this department, and there have been other severe strikes in the history of Shieldhall, perhaps because it is not yet wholly copartnership.

The industries now carried on by the Wholesale are: At Shieldhall, boots, tailoring, shirts, underwear, hosiery, ready-made clothing, cabinet-making, brushes, tinware, printing, preserves, confectionery, tobacco, coffee-essence, pickles; at Glasgow, mantles, water-proofs, aërated water, saddlery, tailoring, and sausages; at Edinboro, one of the finest flour mills in the kingdom, and an oatmeal mill; at Selkirk, tweed works; at Grangemouth, a new and fine soap works, and at Larbert, the Carbrook Mains Farm. The society went into the manufacture of soap because the manufacturers of a popular brand joining the boycott refused to sell to it. The results of these productive operations, exclusive of the farm, were a profit for the half year ending December 25, 1897, of $107,895, on a product for the six months of $3,236,655, an increase of nearly two-thirds over the corresponding period of 1896. From the foundation of the productive works up to June 30, 1897, the total output, as given in the Co-operative Wholesale Societies' Annual for 1898, had been $16,288,000, on which the profit has been $777,738.

At the members' meeting for the third quarter of 1897, the chairman said: "In our productive efforts, the wheels go merrily. Chancelot is sending out six thousand bags of flour every week. Junction mills are turning out five hundred bags of oatmeal weekly. Selkirk is contributing its quota to the general success. The boot and shoe factory is making twelve thousand pairs weekly. The tobacco factory is sending out $12,500 worth weekly. The clothing factories are busy. The printing departments are finding favor everywhere. The cabinet and brush factories are ever increasing, while last, but not least, the new soap works at Grangemouth has become a candidate for your co-operative favor, the goods already produced being of a high class. We can now make from forty to fifty tons per week." For the last half of the year 1897, the net sales of the Wholesale were $11,673,108, an increase of $1,506,515. The society, in 1897, shows the largest increase it has ever had in one year. In the productive departments the sales were $3,236,655, an increase of 57.9 per cent. The profits of all the departments of the Wholesale, distributive and productive, for the half year, after meeting interest, depreciation, and all general charges, were $437,100. Of this, $343,620 were distributed in bonus on purchases and $19,359 in bonus on wages; $24,316 was added to the reserve, and a balance of $49,800 was carried to next year.

Shieldhall provides dining and recreation rooms for its workers. These rooms occupy an entire building in a central position three stories high, with a total floor space of 24,840 square feet. There are separate rooms for men and women. These rooms are supplied with newspapers, magazines, and games, and are in charge of a committee elected by the working people. The recreation rooms are on the first floor; on the two upper floors are the dining-rooms. The food is cooked by steam and gas, and the staff consists of a chef and thirteen assistants. Food is supplied to the workers at cost prices, the dining-rooms being run to pay working expenses only. No employé need take his meals in the dining-room unless he prefers to do so. The average number who eat there is fifteen hundred a day. The gardens and grass plats around the buildings at Shieldhall are being continually encroached upon by the extensions and alterations always going forward. The society has its own building staff, and does its work without the intervention of the contractor, under direction of its own master of works.

I found in all the departments the claim made that the wages paid were the highest demanded by the trades-union, and that in addition to the bonus. It must not be supposed that the Wholesale is in any way qualified to give its workers these advantages in hours, wages, and conditions by any monopoly it has in the co-operative market. Its constituent societies are entirely independent, and buy its goods or not, as they choose. One of the commonest complaints of the co-operative manufacturers, as well as of the co-operative storekeeper, is that they get only a fraction of the custom of the co-operative world.

The earlier advocates of copartnership were hardly prepared to accept the Scottish Wholesale as belonging to their system. They were not willing to regard as a copartnership society a great centralized establishment like this, making its own goods in productive departments. They desired to see co-operative manufacturing undertaken by distinct societies. In a pamphlet on this subject by Thomas Hughes and E. Vansittart Neale, "The Co-operative Faith and Practice," they point out the objections to this centralized system. For one thing, they show how future security is impaired by the fact that the English wholly, and the Scotch to a large extent, pay away the profits of the business in dividends as fast as they are created, while in true copartnership the profits, by their accumulation and conversion into shares, would be an ever-growing fund for the safe conduct of the business. They close their pamphlet with an impressive passage:

"We desire, in thus taking leave, as it were, of a work to which more than forty years of our lives have been devoted, to raise as solemn a protest as we can against the purely commercial spirit by whose insidious growth the fair flower of co-operation is in danger of being blighted. We had hoped that to our country might be reserved the glory of pointing out to other nations the true path from the wilderness in which industry has so long wandered, to the land of industrial peace to which we do not doubt that at some time, under some leaders, a way will be found. But that hope we must abandon for the generation from which we are passing away, if the societies of our Co-operative Union, faithless to the principles to which they are pledged, permit their greatest wholesale center to deal with the claims of labor as if it were an ordinary joint-stock company, by distributing the profits of production according to the wealth of the recipients measured by their purchases, instead of according to the energy manifested in their productive work.

"We hope and pray that it is not yet too late to arrest this evil; that when our thousand associations thoroughly realize that they stand face to face with the momentous question: How shall labor be dealt with by the co-operative capitalist? they will reaffirm in practice the principles to which they have assented in theory, insisting that they shall not only be paraded in congresses, and proclaimed in cooperative gatherings, or other occasions when we desire to figure as the reformers of society, but be embodied in the practice of those great institutions on which, in a large measure, depends the hope of making co-operative action as triumphantly successful in production as it has already become in consumption and exchange."

One of the writers of the Labor Association, Mr. R. Halstead, speaking of the Scottish Wholesale, says the fact of its adopting "some amount of local control in its workshops and factories, proves that in Scotland the working class grip of the democratic principle is stronger than the centralizing tendencies" of the Wholesale. "The Scotch thistle is a wonderful and hardy plant, accustomed to adverse conditions; and the fact that it would yield some genuine co-operative figs, is another instance of the miraculous achievements which follow in the wake of Scotch character and enterprise."

There is another interesting profit-sharing and management-sharing industry in Glasgow, the United Co-operative Baking Society. It occupies very extensive and handsome quarters on McNeil street, and bakes three hundred tons of flour a week. It is purveyor and caterer for the public as well as baker; takes charge of excursions, soirées, supper parties, and marriages. "We furnish everything for weddings," one of them said, "except brides." This society runs a number of restaurants. It has its own farm, and has 103 acres in one piece for the production of roses and flowers for its business. Like the industrial colony at Shieldhall, this bakery is always building. It has not stopped making extensions for years, and has one hundred and fifty men continually at work in its building department.

This society is a federation. Its members are all societies. It began in 1869 with eight of these, with sales of $25,000 and a capital of $1,690. At the end of 1896 it had seventy-six societies; its sales were over one million dollars, with profits of $133,000, and its capital is now about $666,000. The sales for the quarter (thirteen weeks) ending October 30, 1897, were $373,000, an increase over the corresponding quarter (fourteen weeks) of last year of $94,000. There are 903 employés, including the members of the building department. All of the employés belong to the union of their trade, if there is one. Its bakers get one shilling more than the trades-union wages, and their bonus in addition. The hours of work in the bakery are fifty a week. Its women are given the dresses they wear while at work. The rooms are cooled by artificial means, because so many of the girls took cold in the overheated air. Almost all of these women are out of co-operative families. A large number of horses and wagons is kept busy in delivering goods, and one of the features of the buildings is a stable which has three stories of stalls.

This society has always shared its profits with its workingmen, and five and one-half years ago it admitted them to a voice in the management. It formed a society to consist of those employés who choose to become stockholders, and to be entitled, like other member societies, to send representatives to the general meetings. This is called the "Bonus Investment Society." At the general meetings of the employer—the United Co-operative Baking Society—this Bonus Investment Society speaks in the name of the whole body of employés. But the rules of the baking society do not admit any of its employés to places upon the managing committee or to any other office. The balance-sheet for the quarter ending October 30, 1897, shows that of the profit of $54,000, $43,995 was distributed to purchasing members, while $8,495 was paid as bonus to the employés at the rate of two shillings sixpence on the pound of their wages. At the annual social meeting in February, 1898, it was reported that the last quarter of 1897 showed a continuance of previous rapid progress. Sales were $419,166, an increase of $95,697. During 1897, the society used up 2,261 bags of flour per week, and in February this had increased to 2,470, or 310 tons of flour. The number of employés has increased to 911.

It has had not only private, but co-operative competition to struggle with. The Scottish Co-operative Wholesale tried to crush it out on the ground that a co-operative society should itself make everything which its members require, but did not succeed. The ethics of co-operative competition, as laid down by Mr. Jones in the passage quoted above, allows it only against societies "not of the federal type." But here we see the Scottish Wholesale, like the English Wholesale, seeking to destroy a fellow "federal" association.

The bakery has a joint committee on educational work, managed by three directors and three employés, and on its bulletin-board at the entrance to the works I saw posted notices which give an idea of the variety of its duties. There were gospel meetings, a gentlemen's swimming club, a ladies' swimming club, a woman's guild, a literary class, a music class, a draft club, the Bonus Investment Society, physical drill, an ambulance class, and fire brigade. The report of this educational committee for the year ending October 30, 1897, contains one or two passages worth reproducing: "There are over one hundred young people under eighteen years of age in the employment of the society, and recognizing that our responsibility for these young people does not end with a wage payment, or even with the allocation to them of a proportion of the profits earned, we have been trying to gain their confidence and be helpful to them. We had an interesting meeting with them in the Cross Tea Rooms, at which Doctor Dyer, of Glasgow school board; Mr. Campsie, of Kinning Park, and Mr. McGregor, from the committee, kindly attended and gave stimulating addresses, urging the young people to take advantage of the facilities offered by the school board for self-culture."

This committee also goes into political work. Under the head, "Public Action," the report says: "We interested ourselves in the school board election, doing what we could in support of the Progressive candidates. We also gave our hearty support to the Free Libraries proposal. Previous to the subject coming before the city council, we wrote our ward representatives, urging them to support the proposal, and we regret the retrograde decision that the council came to on that occasion."

Action of this kind by co-operative societies is not unusual. In a letter dated March 17, 1898, an official of the Royal Arsenal Society, at Woolwich, writes: "You will be interested to know the society is taking an active part in the municipal elections, now pending, and is nominating candidates in conjunction with the other local trade and labor organizations. This, of course, is only supporting the extension of co-operation from voluntary co-operation to municipal co-operation, or, if you like, from voluntary to compulsory co-operation, and is keeping within the highest principles and ideals of the co-operative movement."

Copartnership has other public aspects. In his testimony before the Royal Commission on Labor, Mr. Henry Vivian, as the representative of the Labor Association, stated that in their view the copartnership of the worker ought to be made a condition by the government in giving franchises to joint-stock companies, and that this principle should also be admitted by the government in its own employ. He said: "We are satisfied that the right of the actual worker in these to a voice in the management, and a share in results, will have to be recognized if the stereotyping of wages and conditions of employment are to be avoided in the future, and struggles between the community and its workers are to be escaped. Strikes do now occur in municipal docks, gas works, etc., as they do in private concerns." He was asked if he would favor copartnership societies being recognized in giving out public contracts, and in answer said: "The city of Paris has already set a precedent in this respect. Preference is given there to contractors who embody the copartnership principle in their works."

Even though the workingman could not be a shareholder in the government industries — as there would be no shares — he would, by the copartnership principle, be given a voice in the management of the government railroad, or armor-plate works, or what not. So when the socialist parodies the crude cry, "The mine for the miners," "The land for the laborers," by adding, "The school for the schoolmasters," "The sewers for the sewer cleaners," labor copartnership promptly accepts the challenge, and admits it would demand for the schoolmasters a direct vote in the policy of the school, and for the sewer cleaners the right to participate in the management of the sewers. But not an exclusive voice.

CHAPTER XVI

"SET ON A HILL"

On the hills of West Killbride, overlooking the sea, in one of the most beautiful spots on the shore of Scotland, stands a handsome light-gray stone building, in which the co-operative movement of Great Britain reaches its highest expression. It is the Scottish Co-operative Convalescent Seaside Home, which has been built by the co-operators' money, as a place of rest and recreation for their tired and sick working people.

Seven years ago Mr. James Deans and Mr. Robert Duncanson, walking through the Glasgow Green, fell to talking of an old co-operator who was getting ill and frail. There ought to be, they felt, a place of refuge for him and such as he. To think of such a thing was, with the co-operators, to do it. Papers were prepared, to be read at conferences and other meetings. The result was the appointment of a committee by the Ayrshire Co-operative Conference Association. In an account which has been printed of the work, it is said that it was felt "the time had come when the co-operative movement in Scotland ought to lift itself above a movement solely for making and dividing profits," and when the societies should begin to devote a portion of their rapidly accruing wealth to getting for themselves "a new and higher class of benefits." "The life, and soul, and glory of the co-operative movement consists," the account continues, "in the ideal it has set before it of ameliorating, if it cannot totally remove, the sufferings, the misery, and the sorrow of the poor struggling masses of the nation, and of making the human life of the poorest of the people contented, sweet, and happy. In proportion as it travels toward this ideal will the brightness of its fame and honor increase; but in proportion as it sets this ideal aside, or delays in trying to realize it, will its glory and prestige, and its claims upon the community decline."

These are fine sentiments, but the finest thing about them is that there was, also, the deed to make them good. It was decided that this ideal could be expressed in no more practical and beautiful way than by the erection of a seaside home, where workingmen and women, recovering from accident or illness, could rest in comfort for a few weeks, and, under the invigorating influences of the sea, grow back to health and strength. The suggestion was well received, and the appeal to the societies for subscriptions to the stock, made by circular and by personal visit from members of the committee, had a favorable response. It was not long before $11,500 had been raised, and this made it possible to secure the site. This was two acres of land in one of the most attractive and finely sheltered spots on the upper part of the Ayrshire coast. The cutting of the first sod for the corner-stone by President Maxwell, of the Scottish Co-operative Wholesale Society, was made a festival, according to the usual procedure of co-operators. As subscriptions did not come in rapidly enough under the organization of the men, the Women's Guild took it up, and, notwithstanding their sex, the men say, did "yeomen's" service. Many of them gave months of their time, going about collecting subscriptions and getting work done for a bazar, in aid of the funds for the building and furnishing of the home. This was held in Glasgow during the New Year's holidays of 1894. The use of the city hall was given by Glasgow, and the bazar was made an exhibition of co-operative products with appropriate entertainments. The affair was most successful, and with the help of redoubled effort on the part of the men, the necessary money was finally raised. The locality of the Seaside Home is a favorite resort for the Glasgow merchants, who have a large number of summer homes here. Many of these, imbued with the hostility of private traders, tried to prevent the co-operators from getting the land to build on, but in this they were unsuccessful. The owner of the site was a woman, and to overcome the prejudices which had been excited in her mind against them, Mr. Deans and two of his associates were sent to ask a personal interview, and show what fine people the co-operators really were. They took great pains to assure the lady that they were not nearly as good as the average. "We are not the pick," they said, "just three of us who happened to be most handy to come." Many of the old-fashioned co-operators looked upon the project as chimerical in the highest degree. Mr. James Deans described to me his encounter with one of this kind. "I was leaving my work early to go to Kilwinning to a meeting to advocate the cause of the Seaside Home, and at the gate of the works I met an old co-operator who said:

" 'Weel, Jimmie, whaur awa' the necht?'

"I told him.

" 'Hoo much are ye wantin' to get?'

"I said we wanted ten thousand pounds. He eyed me up and down—scrutineered me.

" 'Weel, Jimmie, A aye thocht there was a lot of bees bummin' in your bonnet, but A never thocht ye were sae stuipid as that. Ye'll never get twa hunderd poonds.' "

But the spirit of co-operation seems destined always to get what it wants, sooner or later, in one way if not in another. The societies subscribed $35,000, and an old and devoted co-operator, Mr. Barclay, gave $10,000; the $5,000 still needed was borrowed. The architects who built the store of the Co-operative Wholesale in Glasgow, gave their services free. Over the main stairway in the home is a handsome colored-glass window, which was given by the glaziers. In it is commemorated, by portraits, the original committee. The home has now been open two years, and already pays expenses during the summer, but runs behind in winter when there are fewer guests. It is the finest convalescent home in Great Britain, and, as it stands on its hillside looking over the sea, it would be considered a beautiful structure anywhere. It can accommodate eighty or ninety guests. At the time of our visit forty-nine were there, men, women, and children, all co-operators. The manager said with great pride that during the recent fair they had entertained

102, and every one had had a bed and a blanket. The building has four floors, including basement. Four dormitories and other special accommodations are provided for the invalids. There are smoking, library, and recreation rooms, and a mothers' room, with cots and a cooking-range, for sick children. The building is heated by hot water throughout. Besides the housekeeper and manager, there is a staff of eight maids, one girl and one man. The home is for members only of the societies which are shareholders. Each of these shareholding societies takes from the amount of one pound upward and for each pound of stock it owns, has the right to send a member for a fortnight each year at a charge of $2.50 a week. The Kinning Park Co-operative Society, for instance, took £100, and that gives it one hundred tickets for two hundred weeks' entertainment each year. On the doors are such inscriptions as the following, which I copied from one of the bedrooms:

> "THIS ROOM
> "WAS FURNISHED BY
> "CLYDE BANK
> "DISTRICT CO-OPERATIVE
> "EMPLOYES,
> "JUNE, 1896."

They raised for this purpose $75. On the door of the sitting-room appears an inscription that this was furnished by the members of the Women's Guild. This cost $150. One would have to see to appreciate the satisfaction with which the co-operators who were good enough to conceive this plan and fortunate enough to carry it out look upon their work. "We are better satisfied with this," said Mr. Deans, as we went over the building, "than with all our other co-operative work. It is a great advertisement for the co-operative movement. It has made people think. They have seen that co-operation is a generous work as well as a commercial success; that it takes humanity at its weakest point when suffering and seeks to lift it up. It is the finest outcome the co-operative movement of Great Britain has to show. No one comes here as a charity. Every inmate is such by right. This beautiful building in this glorious spot is their home, built with their money. Here they know that they are welcome. Here they are the lords of the manor." We were fortunate, in our visit, to find in the home its Prince Bountiful, the Mr. Barclay who had given it $10,000. He did the honors as host and led us, with joy and pride, through the broad halls and into the fine, light, airy rooms, gay with sunshine and happiness. His had been a typical career. He left his home in Ayrshire, thirty years ago, with threepence in his pocket, and without any large operations, by building each day a little more on the accretions of the thrift of the day before, has put himself in possession of a comfortable little fortune. He has no wife or child. It is easy to see that he has married and adopted the Seaside Home.

The view from the roof to which Mr. Barclay led us is "glorious." Opposite, across the sea, is the Isle of Arran, with Mount Goatfell lifting itself into the sky. The mouth of the Clyde is five or six miles away, and all the foreign commerce of Glasgow sails by the home. Off to the south are the "banks and braes," and "the bregg of Bonnie Doon," of the land of Burns. The spires of Ayr can be seen, and on a fine day the Burns monument can be picked out with a glass. Mr. Barclay saw us to the door, and his honest face bubbled over with human kindness. His farewell words were: "Go, and spread the news of this co-operative movement. It has been a gospel dear to me a great many years."

CHAPTER XVII

BEGINNING AT HOME

When democratic thought sought expression in democratic action in Europe, the Continental workingmen began kicking over thrones and slicing off the heads of kings and queens. The English workingmen saved their tuppences for flour and tea and fustian clubs to buy social regeneration for themselves out of what they could save by this self-help. Penny by penny and sacrifice by sacrifice they have gone on building, tolerated as enthusiasts, until the conventional world wakens to see a very great Fact.

This co-operative movement after fifty years of struggle, has had fifty years of living prosperity, and a greater prosperity is coming into plain view.

It has achieved an economic footing of a hundred millions of property.

Some of the best men and best thinkers have given it leadership.

At least one-seventh of the population of

Great Britain have been enlisted in its ranks—
the pick of the working-classes of the greatest
of European nations.

A program has been evolved looking to
thorough-going social reconstruction.

It is an established religion; for co-operation
is not a method of business merely, but an
ideal of conduct and a theory of human relations. Without cathedrals, creeds, ritual, or
priests, it has not only openly professed, but
has successfully institutionalized the Golden
Rule in business.

As significant as its cold facts and figures
are its recognition of conscience and sentiment
as business factors of the first force, and its success in establishing among men a new method
of union. A movement like this of self-help,
growing of itself, already big and getting
bigger, with a clear and high purpose to do
great things, even the greatest, with so broad a
constitution that it may fairly be said to be of
the people, is a fact with which no one who
wishes to be well informed will neglect to acquaint himself, and no one who wishes to help
on the coming of human brotherhood but
will work for its extension. In contrast
with the unscientific individualism which
preaches that selfish interest is the only actual
and practical motive in the world of values,

we have in co-operation the individualism of millions of men and women organizing millions of capital into successful business under the ethics and economics of each-other-help and all-the-world-help as well as self-help. Here is applied brotherhood; here, the Golden Rule realized; here, a political economy of the kind that seeks wealth for itself by creating wealth for others.

Co-operation has won the right to be accounted the most important social movement of our times outside of politics. It is, of course, only a half truth. It is the voluntary or domestic expression of the same resurgent spirit of self-help by each-other-help of which the republic, democracy, and the hopes of socialism are the political or public expression. Each of these—the private or voluntary, the public or political—is a half truth; but the world needs half truths to make up its whole truth.

The trades-unions have organization, local and federal; the German Socialist Labor Party has votes; but co-operation has organization, federation, votes whenever it shall choose to use them, and—what none of the others have— an economic foothold wide enough to give it a sure place from which to work its lever to move the world. No other social movement of our times surpasses it in radicalism; no other has as many members; no other approaches it in achievements nor solidity of resources in men and money. Labor copartnership is its most advanced element; it has the historic prestige of all the great names of co-operation, the crusading spirit with its accumulating momentum of moral enthusiasm. It is growing as the distributive movement, once fairly started, grew, by leaps and bounds; and it speaks the democratic invitation to all, which in human history has been the only thing that has never been turned back, nor overcome, nor silenced. It is a law of social dynamics that leadership in any movement passes steadily into the hands of its radicals. If this proves true in co-operation, the head of the column will be taken by labor copartnership

The British co-operative movement is a living reality, and is in the flooding tide of growth, success and confidence. This tide is not yet at the full. It is still rising. Between the time of making my notes in England and correcting my proofs a few months later, the returns which came in made it necessary to increase nearly all my figures. Both in England and Scotland the productive and distributive concerns, with very few exceptions—like those of Walsall and Burnley—are making new records of success. The English Wholesale passes, for the first time in its history, the $15,000,000 mark of business in one quarter, and the annual turnover of its bank becomes $200,000,000. The Scottish Wholesale shows a gain of nearly sixty per cent in its workshops. The Leicester Equity boot and shoe works add to their capacity seventy-five per cent. In Ireland in the year the number of creameries rises from 94 to 131, and of people's banks from 3 to 13. The statistics in the appendix show the march. The movement is pressing forward into fresh enterprises in old fields and ventures in new worlds like farming. The renaissance of the labor copartnership ideas of the founders evidences that there is a moral expansion as well as a material.

How can the reader learn more? If possible, go to England and study co-operation on the ground. One hour of seeing is worth a year of reading—and seeing is believing. There are no books which give the facts to date, though Mr. George Jacob Holyoake's "Jubilee History" of the Leeds co-operative store from 1847 to 1897 is a record of the most successful local store in Great Britain, which illuminates the whole movement. This can be procured for one shilling sixpence from Mr. John W. Fawcett, secretary of the co-operative store, Leeds, England, or from the Co-operative

Union, Manchester. For those who cannot make personal investigation, the only recourse is to correspondence and the current periodicals and documents of the movement. Mr. J. C. Gray, secretary of the Co-operative Union, Long Millgate, Manchester, England, can furnish information and lists of tracts, documents, and other co-operative publications, but not those of the copartnership school. For news of Scotland, inquire of Mr. James Deans, 20 Moodie's Court, off Argyle street, Glasgow; for the Labor Association, of Mr. Henry Vivian, 15 Southampton Row, London, W.C.; for the Productive Federation, of Mr. Thomas Blandford, 19 Southampton Row, London, W.C., England; for the Irish Agricultural Organization Society, of Mr. R. A. Anderson, Secretary, 22 Lincoln Place, Dublin. The secretary of any co-operative society would answer inquiries. Any one who wants to keep himself informed in regard to the movement should subscribe, by international money-orders, for its periodicals. These are:

Labor Copartnership, published monthly by the Labor Association, the organ of the labor copartnership movement, at 15 Southampton Row, London, W. C.; price, one shilling sixpence a year.

The Co-operative News, published weekly,

at a penny, by the Co-operative Union, Long Millgate, Manchester.

The Irish Homestead, published weekly, at a penny, by the Irish Co-operative Newspaper Society, 22 Lincoln Place, Dublin. This is the organ of the Irish Agricultural Organization Society.

The Agricultural Economist, a monthly, edited by Mr. E. O. Greening, author of "The Co-operative Traveler," five shillings a year, 3 Agar street, Strand, London, W.C.

Can we do anything like this in America? Excepting building societies and mutual insurance, co-operation here has been mainly distributive with no such signal successes as the great stores of Great Britain to boast. The Co-operative Union of America, of Cambridge, Mass., is a federation of societies to promote here "the Rochdale plan, devoted to the cause of distributive co-operation," and it publishes, at fifty cents a year, *The American Co-operative News*. Of co-operative production, outside such enterprises as the co-operative creameries east and west, there is little; and of production of the co-partnership sort, practically nothing, except so far as the Labor Exchanges are under that head. But in all these directions, much faithful, earnest pioneering is being done, the indispensable preliminary, as foreign experience shows, of success. I have written the story of Labor Co-partnership in the hope that it might give encouragement to the co-operative spirit of this country, none the less if our lines of progress should have to differentiate themselves from those taken abroad.

We are a different people from the English, and our circumstances are different. When the English co-operative movement struck its present roots, the consolidation of industry had hardly begun; while to-day, in America, it does not seem to have much farther to go to bring us to the consummation of an industrial feudalism which will make impossible any individual, or even co-operative initiative or independence. Many of the American students and workers feel faintheartedly that the development of the trusts makes it hopeless for co-operation to obtain that foothold here which the English workmen succeeded in obtaining a generation ago. There is, too, the feeling among many of us that Americans are not a co-operative people; that, as has been said, our genius is more for co-talking than for co-working. But this is the greatest of mistakes. America has been doing a co-operative work even more remarkable than that of England. Each nation has followed its genius on the lines of greatest need and least resistance. English co-operation has been industrial; ours has been political. The achievement of America in uniting in one common life and one co-operative citizenship the African and European, and even Asiatic types, which elsewhere glare at each other with hatred across frontiers of bayonets, is the greatest triumph of co-operation which the history of civilization has yet shown. Compared with this, the English task of getting the men of one trade and one locality to unite in so simple a social function as purchasing or making commodities, was easy. As America advances from political co-operation to industrial, its accomplishment is likely to be as much grander than that of Great Britain as the task is more difficult on account of extent of territory, racial variance, and the newness of social life. There is some reason to expect that the American evolution in co-operation will be partly along the lines of colonization and the establishment of new communities. The American is a colonizer. All Americans are colonists from abroad. The West was colonized from the East. We colonized California, and are now colonizing Alaska. One hears to-day in every direction in America of plans for co-operative colonies. That aptitude for political co-operation which has been forced upon us by the necessity of amalgamating so many nationalities fits us specially for community building. Village communities so placed as to have an easy command of the necessities of life can achieve an independence of the railroads and trusts, and an opportunity for the members for production and exchange with each other which would be out of reach if they remained disconnected individuals, lost in the meshes of the highly complicated and centralized economic system of the American business world. The English co-operators, I found, look with great hope to America. Thomas Hughes, in one of the letters which I saw, said in 1887: "I should not be surprised if America took the lead out of our hands." Mr. E. O. Greening, one of the closest of living students of the co-operative movement, said to me: "Co-operation will be slow to take root in America, but once started, it will develop into forms far greater than anything here." Mr. George Jacob Holyoake made an interesting suggestion: "The churches," he said, "may make co-operation go in the United States." Rev. F. D. Maurice pointed out to the workingmen of Great Britain that the one indispensability for the success of co-operation was mutual confidence. Certain is it that co-operation in America must progress along lines where the people know each other, in their trades-unions or some other organizations—perhaps, as Mr. Holyoake has said, in the churches—or where they will be forced together as in village communities. There is hardly, as yet, anywhere in America that neighborhood life which made it easy for the weavers of Rochdale or the shoemakers of Kettering, who had known each other almost from childhood, to get together. In London, where the population is nearly as shifting as in America, it has been found almost impossible to acclimatize co-operation. Co-operation can go on in America as in Great Britain and on the Continent, only by the help of men of means, culture, and good-will—men well-to-do in good deeds as well as title deeds; men like Horace Plunkett, who is leading the Irish in co-operation to-day, and like Holyoake, Ludlow, Hughes, Neale, Kingsley, Maurice, Owen, Godin, Leclaire, Van Marken, Schultze-Delitzsch, Raffeisen, and Luzzatti, who have been leading European workingmen. What co-operation needs here, as elsewhere, is not philanthropy, but leadership; not endowment, but initiative.

APPENDIX

I—LABOR COPARTNERSHIP SOCIETIES.

The Labor Association publishes the following table of the societies which are included in its summary of the statistics of copartnership given on page 227. These statistics relate only to copartnership in co-operative productive societies. Copartnership of labor in co-operative distributive societies is not touched, nor is the important growth of copartnership in businesses of a capitalistic origin:

ESTABLISHED NAME.	Capital Share Loan Reserve.	Trade.	Profit.	Dividend on Wages.	
				Am't.	Per £ of Wages.
ENGLAND.					
TEXTILES, ETC.	£	£	£	£	s. d.
1860 Eccles Manufacturing.....	21,346	19,065	952		
1870 Hebden Bridge Fustian....	36,825	46,862	4,638	758	1 0
1872 Airedale Manufacturing...	7,322	19,047	607	19	0 3
1874 Leek Silk Twist...........	3,862	12,230	321	168	1 6
1876 Leicester Hosiery.........	31,785	48,885	2,602	190	0 4
1886 Burnley Self-Help	22,420	78,523			
1886 W. Thomson Sons, Ltd ...	23,168	31,180	2,340	611	1 9
1886 Macclesfield Silk...........	23,557	29,902			
1888 Nelson Self-Help	3,188	29,029	547		
1890 Delph Woolen	1,380	1,241	39		
1891 Sheffield Tailors...........	304	531			
1892 Nottingham Tailors.......	351	832			
1892 Co-op.Inst. Tailoring Dept.	686	1,143			
1893 Kettering Clothing	5,970	18,894	1,370	375	1 4½
1894 Macclesfield Fustian.......	984	2,071			
AGRICULTURAL, ETC.					
1867 Agricultural, etc., Assoc...	51,331	62,503	542	139	
1883 Assington Farm...........	3,676	1,031			

ESTABLISHED NAME.	Capital Share Loan Reserve	Trade.	Profit.	Dividend on Wages.	
				Am't.	Per £ of Wages.
	£	£	£	£	s. d.
BOOTS, LEATHER, ETC.					
1881 Northamptonshire (Wool st'n)	1,986	5,534			
1884 Bozeat	884	2,030			
1885 Norwich	484	2,654	102	14	0 8
1887 Leicester ("Equity")	25,090	47,296	2,390	740	0 11
1888 Kettering	9,328	29,204	2,233	724	1 7½
1889 Bristol Pioneers	822	1,267	29		
1890 Hinckley	2,708	6,568	109	20	0 2½
1890 Nantwich	970	4,101	174		
1891 Barwell	2,731	6,322	410	120	1 2
*1891 Walsall Horse Collar	232	1,161	52		
1891 Street Boot	326	57			
1892 Anchor (Leicester)	1,119	4,420	150	24	0 3
1892 Glenfield	967	5,354	428	100	1 4¼
1892 Higham Ferrers	573	4,377	129		
1892 London Leather Manf't'rs.	2,963	8,702	178	51	0 3½
1892 Rushden Trade Boot	322	6,288			
1893 Desborough	1,545	15,723	*96	*31	0 *9
1893 Rothwell	1,165	9,432	355	73	0 6
1894 Penken (Leather Productive)	2,195	11,649	3		
1895 Canterbury Tanners	1,219	584	29		
1896 Kettering Union Boot	519	1,016			
1896 Leicester "Self-Help" Boot	198	957			
METAL TRADES.					
1873 Walsall Padlock	6,996	16,481	1,103	710	2 0
1873 Sheffield Cutlery	1,416	1,743	221	22	1 6
1876 Coventry Watch	2,338	4,003	325	78	1 3
1885 Keighley Iron Works	6,824	7,109	569	41	0 5
*1887 Midland Tinplate	932	3,956	162		
1888 Alcester Needle	3,690	2,738			
1888 Bromsgrove Nail	546	1,571	149	30	0 9
1888 Dudley Bucket and Fender	3,836	14,371	1,155	329	2 0
*1889 Cradley Sheet Ironworkers	1,616	6,039	64		
1892 Calderdale Clog Sundries	838	2,374	115	13	0 6
1893 Leicester Engineers	532	1,274	30		
1894 Sheffield Sheep-Shear	2,823	*4,649	454	240	1 0
†1894 General Engineers	577	300			
BUILDING, WOOD, ETC.					
1873 Household Furnishing, Newcastle	5,432	12,510	312		
1888 Brixton Builders	772	8,516	683	205	0 9
1889 Oxford Builders	177	666	46	9	0 9
*1889 London Cabinet Makers	201	82	3		
1891 Bradford Cabinet Makers	4,153	5,043			
*1891 Chelsea Decorators	35	11	2		
1891 General Builders	1,765	1,384			
1892 Bolton Cabinet Makers	671	3,844			
1892 Newcastle Cabinet Makers	199	2,275			
1892 Medway Barge Builders	877	360	23		

*No return, for 1896 figures are the latest available.
†Estimated.

APPENDIX

ESTABLISHED NAME.	Capital Share Loan Reserve.	Trade.	Profit.	Dividend on Wages. Am't.	Per £ of Wages.
	£	£	£	£	s. d.
1893 Plymouth House Painters	45	178	21	4	0 9½
1894 Kettering Builders	1,954	8,032	670	234	2 0
1894 Sheffield House Painters	119	964			
1896 Cambridge Builders	91	431	39	8	1 0
PRINTING, ETC.					
1869 Manchester Printers	39,838	67,309	3,639	570	0 6
1885 London Bookbinders	416	845	49		
1892 Leicester Printers	1,395	2,659	195	44	0 8½
1895 Nottingham Printers	308	553	45	5	0 5
1896 Blackpool Printers	817	551			
VARIOUS.					
1816 Sheerness Economical	17,911	28,569	3,407	39	0 7
1885 Co-operative Sundries	8,879	22,405	1,269	140	1 3
1887 London Productive	2,725	2,850	44		
1889 Bass Dressers	1,170	4,238	143	52	0 11
1892 Brownfield's Guild Pottery	24,643	*21,739			
1892 Kent Brickmakers	4,837	3,741	261		
1892 Nottingham Bakers	205	1,005	14		
1892 Manchester Mat	871	635			
1892 Bristol Piano	1,325	868			
1896 Typewriters	54	73	5		
Eighty-four English Societies	451,350	836,705	35,240	6,939	0 0
SCOTLAND.					
1840 Kettle Baking	1,323	4,006	620	420	2 1
1862 Paisley Weaving	57,981	71,497	4,330	515	0 10
1868 Scotch Wholesale (Productive Departments)	355,345	720,743	39,484	†4,200	0 7¾
1868 United Baking, Glasgow	133,252	220,536	27,490	3,698	1 10½
1873 Edinboro' Printing Co	19,495	10,737	847	153	0 10¼
1886 Scottish Farming	14,045	13,007	263		
1892 Condorrat Quarrying	550	2,521	147	102	1 3¾
Seven Scottish Societies	581,991	1,043,047	73,181	9,088	
IRELAND.					
1889 Sixteen Dairy Societies to 1896 showing division of profits	13,774	109,111	2,350	56	Nil to 2 0
Forty-two others not showing division of profits	29,499	174,327	2,195		
1895 Dalkey Embroidery, etc	644	1,267	25		
1896 Roscommon Pig Feeders	458	345			
Sixty-one Irish Societies	44,375	285,050	4,570	56	
UNITED KINGDOM.					
152 Societies	1,077,716	2,164,802	112,991	16,038	0 0

There were losses for the year of £11,329 by the English societies and £741 by the Irish societies, and none by the Scotch societies—a total of £12,070. The table in LABOR COPARTNERSHIP does not tell how these losses are distributed among the various societies.

LABOR COPARTNERSHIP points out that the real profit to labor—the money profit, is at least double the £16,083 shown here. (See page 229, above.) It adds: "But the money profit is, after all, a small item in the matter. It is rather because of the profit to character and to life that we rejoice to be able to show these increases from year to year."

2—A LABOR COPARTNERSHIP INSTANCE.—GROWTH OF THE KETTERING CO-OPERATIVE BOOT AND SHOE SOCIETY EACH HALF YEAR.

Half Years.	Members.	Capital.	Trade.	Profits.	Reserve Fund.
		£	£	£	£
1889	195	525	2,050	111	22
1889	208	1,033	3,537	225	40
1890	232	1,360	4,614	321	66
1890	271	1,890	5,316	363	116
1891	298	2,351	6,651	456	164
1891	303	2,582	7,222	473	199
1892	358	3,083	10,249	697	219
1892	378	3,582	9,006	667	234
1893	408	4,032	10,920	842	236
1893	433	4,526	8,873	698	270
1894	449	4,839	11,102	857	320
1894	478	5,355	11,180	784	368
1895	504	6,055	12,433	971	400
1895	580	6,993	13,822	940	454
1896	593	7,830	15,585	1,050	520
1896	651	8,677	13,619	951	581
1897	684	9,721	18,008	1,188	656
1897	710	10,536	14,916	972	705

3.—CO-OPERATIVE SOCIETIES IN THE UNITED KINGDOM.

Statistics showing the position and progress of the co-operative movement from 1862 to 1895. (From the Co-operative Wholesale Societies' Annual for 1898. Pounds sterling changed to dollars at the rate of $5 to £1.)

CO-OPERATION IN THE UNITED KINGDOM DURING 1885 AND 1895.

	1885.	1895.	Increase per cent.
Number Societies making returns.	1,441	1,966	36
Number Members.	850,659	1,430,340	68
Capital (share and loan)	$ 55,785,465	$106,656,995	91
Sales	156,529,550	275,501,245	76
Profits	14,943,450	26,945,355	80
Profits devoted to Education	103,560	207,455	100

APPENDIX 341

4—THIRTY YEARS' WHOLESALE DISTRIBUTION IN SCOTLAND.—SCOTTISH CO-OPERATIVE WHOLESALE SOCIETY, LIMITED.

(From the Co-operative Wholesale Societies' Annual for 1898. Pounds sterling changed to dollars at $5 for £1.)

Years.	Weeks.	Capital.	Sales.	Profits.	Weeks.	Years.
1868	13	$ 8,975	$ 48,485	$ 240	13	1868
1869	52	25,875	405,470	6,520	52	1869
1870	50	62,715	526,245	12,095	50	1870
1871	52	90,045	813,290	20,655	52	1871
1872	52	154,655	1,312,650	27,175	52	1872
1873	52	252,165	1,922,445	37,230	52	1873
1874	52	244,910	2,049,735	37,765	52	1874
1875	52	283,755	2,150,845	41,165	52	1875
1876	51	336,095	2,287,645	44,180	51	1876
1877	52	362,840	2,946,105	54,625	52	1877
1878	52	415,870	3,002,950	59,845	52	1878
1879	52	465,385	3,150,485	74,945	52	1879
1880	52	550,895	4,226,105	103,425	52	1880
1881	54	678,565	4,933,230	119,005	54	1881
1882	52	847,145	5,502,940	116,100	52	1882
1883	52	976,980	6,265,770	141,830	52	1883
1884	52	1,220,930	6,501,655	147,175	52	1884
1885	52	1,444,730	7,191,100	198,205	52	1885
1886	60	1,668,265	9,285,760	251,990	60	1886
1887	53	1,836,545	9,050,075	236,390	53	1887
1888	52	2,048,340	9,819,265	267,690	52	1888
1889	52	2,403,110	11,368,910	308,780	52	1889
1890	52	2,876,610	12,378,005	382,725	52	1890
1891	52	3,355,540	14,140,180	445,450	52	1891
1892	53	3,892,470	15,523,840	460,135	53	1892
1893	52	4,348,780	15,677,810	445,580	52	1893
1894	52	4,704,175	15,282,910	442,260	52	1894
1895	52	5,671,345	17,247,305	661,870	52	1895
1896	52	6,186,585	19,112,900	874,910	52	1896
1897	52	6,433,805	22,029,265	860,275	52	1897
Totals.......		$6,433,805	$226,153,450	$6,906,160	Totals.......	

Commenced September, 1868.

5—THIRTY-FOUR YEARS' PROGRESS OF THE ENGLISH CO-OPERATIVE WHOLESALE SOCIETY, LIMITED.

(From the Co-operative Wholesale Societies' Annual for 1898. Pounds sterling changed to dollars at the rate of $5 for £1.)

Years.	Sales.	Years.	Sales.
1864, 30 weeks	$ 259,285	1869.....................	$2,536,085
1865.................	603,770	1870, 53 weeks	3,388,670
1866.................	877,445	1871.....................	3,793,820
1867, 65 weeks	1,658,720	1872.....................	5,765,660
1868.................	2,061,200	1873.....................	8,184,750

Years.	Sales.	Years.	Sales.
1874	$ 9,824,145	1886	$26,115,895
1875	11,230,975	1887	28,566,175
1876, 53 weeks	13,486,830	1888	31,000,370
1877	14,135,260	1889, 53 weeks	35,144,720
1878	13,528,125	1890	37,145,365
1879, 50 weeks	13,226,655	1891	43,832,150
1880	16,698,405	1892	46,504,520
1881	17,870,475	1893	47,630,835
1882	20,191,190	1894	47,219,690
1883	22,734,445	1895, 53 weeks	50,709,585
1884, 53 weeks	23,376,855	1896	55,575,280
1885	23,965,755	1897	59,600,715

Total sales in the thirty-four years, 1864 to 1897............ 738,449,820
Total profits in the thirty-four years, 1864 to 1897............ 9,891,415

6—STATISTICAL POSITION OF THE ENGLISH CO-OPERATIVE WHOLESALE SOCIETY, LIMITED, DECEMBER 25, 1897.

Number of Societies holding shares................................. 1,046
Number of Members belonging to shareholders............... 1,053,564
Share Capital... $ 3,643,745
Loans and Deposits... 6,271,595
Reserve Fund—Trade and Bank................................. 549,415
Insurance Fund.. 1,753,735
Sales for the year 1897... 59,600,715
Net Profits for 1897... 677,805

7.—TRADES-UNIONISTS AND LABOR COPARTNERSHIP.

At a conference of delegates from some fifty trades-unions and kindred bodies held under the auspices of the Labor association in London, March 20, 1898, the following resolution was adopted:

"That this conference of trade-unionists, fully conscious of the need of a more equitable and satisfactory industrial system, declares in favor of the principle of Labor copartnerships as advocated by the Labor association, convinced that it affords a practical means of emancipating the workers from the evils of the wage system, giving them a real share of the profits, responsibilities and control of their labor; and it earnestly calls upon all London trade-unionists to promote cooperative production in every possible way."

8.—ELIGIBILITY OF EMPLOYÉS AS OFFICERS.

The "model rules" which have been prepared by the Labor Association for the guidance of those who desire to organize copartnership works, stipulate for the eligibility of employés as directors, and also enforce the rule of one individual, one vote. They also provide that all employés shall become stockholders, and their share of profits be withheld and accumulated until the stock has been paid for. The societies we have described at Leicester and Kettering substantially follow these rules, as do many of the others. For instance, the rules of the London Leather Manufacturing Society say:

"VII. General Rules 75 and 90.—The Committee of Management shall consist of the President, the Secretary, and seven committeemen, two of whom shall retire at each ordinary business meeting. The Secretary shall retire and be elected at the January meeting.

"VIII.—General Rule 78.—The employés of the society are entitled to elect three committeemen, and are also eligible to any office in the society."

In some of the copartnership societies, the governing committees are drawn from outside shareholders. In others they consist partly of outsiders and partly of workers from the shops; and some are governed by committees consisting of working shareholders employed in the place. And under all these systems the workers are able to exercise a strong influence over the conduct of their own affairs.

A CHARGE OF SWEATING.

Discredit is thrown on the labor copartnership movement by charges that some of the societies tolerate sweating. The Walsall Cooperative Padlock Society is among those found fault with as giving work to small masters outside. It is an interesting illustration of the ability of workingmen, by co-operative methods, to lift themselves out of the misery of a badly "sweated" trade into independence.

It was started during a strike in 1872 by padlock-makers, who were being ground down to starvation. Their society now holds the first place in its branch of the trade. It conducts the most extensive business of the kind in its town, and does it in a large and healthful factory, and it has set the example which will put an end to the deplorable conditions at Walsall, if the private manufacturers will follow. The society began in 1873, with a capital of $415. It now has a factory which cost $12,500, and employs two hundred men. It finds its largest market in India, and the plague and famine cut its sales for 1897 down to $66,300. But the management, notwithstanding that the profit for the year is only $1,105, a diminution of $4,250, in their annual report "suggest that you place at our disposal for charitable purposes during the year £10 ($50), to be taken from the reserve fund."

It could not reform everything at once, and for a time continued to give out some work to be done at home. But the society has recently carried forward its program of putting an end to home work by bringing key filing into the factory. "This is a gain in comfort and health to the workers," the annual report says, but a "monetary loss" to the society.

These co-partnership padlock-makers are defended by the Labor Association as having probably done more to improve the worker's position than any other productive society in the country.

STATISTICS OF 1897.

The London *Daily Chronicle* of May 31, 1898, in its report of the Co-operative Congress, sitting at Peterborough, gives this summary of the figures for 1897 of the Central Board of the Co-operative Union: The number of societies in 1897 was 1845, compared with 1741 in 1896. Of these 1686 made returns showing a membership of 1,591,455 in 1897, and 1665 made returns in 1896, showing a membership of 1,492,371. The sales in 1897 reached £62,287,058, and in 1896, £57,318,426. The profits in 1897 were £6,717,876, and in 1896, £6,337,490. Twenty-seven new distributive retail societies had been registered during the year, and twenty-seven had ceased to exist.

INDEX

Ackland, Joseph, 280.
Advertising, 269.
Agricultural Economist, 332.
Agricultural laborers, wages, 11, 23, 38, 50, 70.
   and co-operation, 82, 85.
   cottages, 70.
   and education, 82, 85.
Agriculture, state aid to, 73.
Allan, John, 255.
America, co-operation in, 124, 332.
   Co-operative Union of, 332.
Anderson, R. A., 69, 228, 331.
Assington farm, 23, 24, 26.
Attendance committee, 177.

Bakery, co-operative, Glasgow, 47, 311.
Ballard, Edmund, 126.
Ballard, F., 133, 136, 145, 149.
Ballard, W., 140, 142.
Ballinagore society, 68.
Bamford, A., 147.
Bank, Co-operative Wholesale, 1, 276.
   Penny, 129.
   People's, 7, 52, 64, 242.
Blandford, Thomas, 78, 331, 240.
Board of Works, Ireland, 68.
Bonus (see dividend on wages).
Bonus Investment Society, 47, 313.
Bonus Loan Fund, 299.
Bonus, sliding scale, 196.
   by South Metropolitan Gas Co., 200.
Bookbinders, 236.
Bookkeeping, co-operative, 19, 29, 91, 129.

Borgeaud, 15.
Boycott, 106, 139, 151, 168, 236, 251, 270, 276.
Brassey, T. A., 231, 236.
Brown, Sir B. C., 259.
Builders, Kettering, 142, 145.
   Leicester, co-operative, 111, 287.
   London, 147, 237.
Building societies, 111.
   American, 332.
Burnley, 154, 161.
Burns, John, 194.
Business, workingmen in, 84, 109, 152, 251, 302.
Butcher, John, 103.
Butchers, co-operative, 252.
Butter, marketing, 92.
Buying and selling societies, Irish, 61.

Cain, J. J., 92.
Campbell, William, 22, 37, 81, 86, 94.
Canadian creamery socialism, 75.
Capital, co-operative, 12, 91, 258, 260, 271.
   sharing profit, 2, 222, 223.
   surplus of, 109, 122, 127, 267, 282, 287.
Capital, withdrawable, 30, 269.
Capitalists repelled by co-operators, 146.
Capitalization of bonus, 108, 115.
   compulsory, 39, 50, 66, 108, 115, 201, 215, 291.
Cattle auctions, 251.
Celebrations, co-operative, 36, 113, 139, 289, 320.

Centralization of English Wholesale, 280, 309.
Scottish, 309.
Charity, 110, 138, 189.
Children, 139, 155, 250, 315.
Christian Socialist, 216, 278.
Christian Socialists, 4, 178, 215, 277.
Church democracy, 15, 16.
Churches and co-operation, 335.
Coal-miners' lockout, 81.
Cocoa works, 237.
Colonization, 334.
Committee, attendance, 177.
compensation, 119.
Community, co-operative, 7, 94, 97, 98, 215, 334.
Competition among co-operators, 70, 116, 138, 237, 278, 302, 314.
Compulsory capitalization, 39, 50, 66, 108, 115, 201, 215, 291.
Comradeship, 32.
Conciliation board, 108.
Congregationalists, 15.
Congress, International Co-operative, Delft, 5.
Paris, 280.
Consumer sharing profit, 2, 110, 167, 222, 234, 307, 313.
Co-operation and agricultural laborers, 82, 86.
in America, 124, 335.
distributive, 1, 18, 264, 332, 340.
distributive opposing productive, 122, 155.
early, 214.
economic superiority of, 187, 266, 271.
future possibilities of, 265.
half-truth, 328.
and landlords, 14, 44, 49.
in London, 235.
publicity, 268.
remuneration of managers, 261, 272.
in Scotland, 275.
and socialism, 185.

statement of A. Greenwood, 294.
subsidized, 274, 242.
Co-operative butchers, 252.
competition, 70, 116, 138, 237, 278, 302, 314.
festivals, 242, 280.
market, 269.
music, 243.
playground, 247.
population, 1, 130.
societies in 1829, 214.
visits, 248.
Co-operative News, 45, 209, 332.
Co-operative Productive Federation, 240.
Union, 9, 164, 169, 219, 254, 265, 270, 331.
of America, 332.
Co-operators, indifference of, 39, 40, 137, 141.
Co-operators and parson, 48, 49.
in politics, 13, 91, 315, 316.
and railroads, 15, 77, 95, 253.
Corset factory, Kettering, 147.
Cottages for Irish farm-laborers, 70.
Craig, E. T., 86.
Creameries, Canada, 75.
Ireland, 52, 56, 63, 70, 71, 72.
experimental, 60.
Crystal Palace Gas Co., 199.

Dalkey, 66.
Deans, James, 42, 169, 250, 252, 302, 318, 331.
Delft, 5.
Democracy, 234, 265, 328.
and the church, 15, 16.
Department of Agriculture for Ireland, 72.
Depreciation in co-operative bookkeeping, 19.
Desborough, 130.
Dining-room, co-operative, 186, 307, 308.
Dividend, sliding scale, 193.
on wages, 30, 31, 38, 39, 47, 66,

INDEX

67, 85, 104, 107, 110, 111, 115, 119, 137, 143, 147, 158, 167, 168, 169, 173, 178, 184, 185, 191, 215, 220, 222, 226, 229, 239, 276, 290, 299, 307, 312, 313.
Donaghpatrick, 61.
Doneraile, 70.
   creamery, 52.
   People's Bank, 52, 64, 242.
Duncanson, Robert, 318.
Dunfermline farm, 20, 26.
Durham, Bishop of, 231.

Economic Review, 215, 280.
Economic superiority of co-operation, 187, 266, 271.
Education, 30, 46, 49, 60, 82, 107, 111, 137, 138, 148, 155, 275, 283, 284, 314.
Eight hours day, 135, 136, 137, 143, 192, 236.
Eligibility of workmen as officers (see employés).
Employés, eligibility as officers, 115, 116, 118, 119, 167, 184, 185, 191, 220, 291, 292, 313, 342.
Employé shareholders, 39, 46, 47, 50, 66, 108, 113, 115, 146, 168, 173, 180, 185, 191, 200, 220, 224, 236, 237, 291, 292, 313.
Engineers, Leicester, 118, 142.
English Co-operative Wholesale Society, 1, 11, 70, 100, 101, 116, 161, 237, 243, 273, 329, 341, 342.
   and employés, 11, 281.
   bonus, 105.
   competition with other co-operators, 70, 117, 279.
   excluded from co-operative festivals, 280.
   Irish creameries, 71, 72.
   production, 102.
   profit-sharing, 276.
   shoe works, 103.
"Equity" shoe works, 100, 106, 132, 276, 285, 330.

Ethics, 65, 123, 131, 137, 150, 188, 189, 250, 270, 327.
Experimental creameries, 60.

Failures, 17, 95, 123, 214 et seq.
Farm, Assington, 23, 24, 26.
   Dunfermline, 20, 26.
   Ipswich, 23.
   Kettering, 13, 47, 48, 128.
   Lincoln, 20, 21, 25.
   Newcastle, 22.
   Scottish Co-operative Farming Association, 20, 41.
   Scottish Wholesale, 19, 20, 26, 47.
   Woolwich, 19, 20, 28.
Farming, co-operative, 9.
   superiority of, 91, 93.
   failures, 95.
   in Europe, 96.
   losses, 18, 41.
   profits, 18.
Fawcett, John W., 330.
Fawcett, Professor, 295.
Federation, 102, 226, 232, 240, 274.
Festivals, co-operative, 242, 280.
Fines, 108.
Framemakers' and Gilders' Association, 262.

Gas Workers' Union, 191, 211.
General Builders, 237.
Gill, T. P., 264.
Gladstone, W. E., 154.
Glasgow bakery, 47, 311.
Godin, 4, 336.
Golden Rule, 3, 250, 328.
Govan Old Victualing, 214.
Government employ and labor co-partnership, 316.
Government loans to individuals, 68.
Grange creamery, 52, 70.
Gray, J. C., 255, 331.
Greening, E. O., 41, 86, 105, 123, 181, 219, 244, 271, 332, 335.
Greenwood, Abraham, 294.
Greenwood, Joseph, 171, 316.

Grey, Earl, 231, 287.
Gurney, Sybella, 223, 230.

Half-truth of co-operation, 328.
Halls, co-operative, 148.
Halstead, R., 282, 311.
Hebden Bridge, 121, 166.
Heckmondwike, 267.
Help one another, 3, 142, 150.
Hodgson, Dr., 295.
Hole, James, 98.
Holiday docking, 237.
Holiday pay, 192.
Holmes, James, 114.
Holyoake, G. J., 4, 80, 98, 109, 124, 169, 219, 250, 262, 269, 271, 276, 280, 285, 294, 295, 297, 330, 335, 336.
Home Industries' Societies, 66, 67, 68.
Hood, Lord, 49.
Hours, 23, 104, 108, 135-137, 143, 192, 229, 236, 238, 302, 312 (see eight hours).
Houses, co-operative, 112, 127, 136, 332.
Huddersfield co-operative store, 269.
Hughes, Thomas, 4, 170, 216, 278, 282, 297, 309, 335, 336.

Indifference of co-operators, 39, 40, 137, 141.
Individualism, 327.
Iniskillen creamery, 71.
Insurance, co-operative, American, 332.
Interest, reduction of, 146, 178, 179.
International Congress, Delft, 5.
Paris, 280.
International Co-operative Alliance, 287.
Ipswich farm, 23.
Ireland, 52, 339.
 co-operative creamery statistics, 59.
 creameries, 52, 56, 63, 70, 71, 72.
 department of agriculture, 72.

home industries' societies, 66, 67, 68.
labor copartnership, 228.
people's banks, 7, 52, 64, 242.
railroads, 77.
saving by co-operation, 62, 63, 77.
self-help, 59.
Irish Agricultural Organization Society, 54, 58, 64, 71, 77, 228.
 affiliated societies, 69.
 Agricultural Wholesale Society, 57.
 Co-operative Agency Society, 57.
Irish Homestead, 71, 230, 332.
Irish land agitation, 46.
Irlam, 281.

Jessop, L., 144, 149.
Joint committee, workmen and officials, 207.
Jones, Benjamin, 103, 104, 160, 168, 262, 273, 279.
Lloyd, 98.

Kettering, 40, 49, 100, 121, 267.
 boot and shoe society, 132, 223, 340.
 builders, 142, 145.
 clothing, 139.
 co-operative store, 48, 121.
 corset factory, 147.
 farm, 13, 47, 48, 128.
Kingsley, Charles, 4, 216, 297, 336.

Labor Association, 9, 220.
 statistics, 2, 227, 230.
 a voluntary organization, 226, 230.
Labor Copartnership, 67, 119, 198, 199, 224, 230, 242, 331.
Labor copartnership in farming, 21, 24, 26, 28, 41, 45, 47, 48, 52, 69, 85, 99.
 in government employ, 316.
 in public contracts, 317.
 statistics, 2, 227, 228, 230, 337.
 ultimate goal, 231.
 exchanges, 332.
 notes, 88.

INDEX 349

Laborer, agricultural, and co-operation (see agricultural).
Land and co-operation, 10, 13, 16, 99.
   out of cultivation, 80.
   decline in value, 80.
Landlords and co-operation, 14, 44, 49.
Laundry, co-operative, 168.
   no bonus, 169.
Leases, farming, 14, 43, 45, 95.
Leather society, 237.
Leclaire, 4, 336.
Leeds co-operative store, 262, 269, 330.
   society, industrial city, 98.
Legacy to Burnley workers, 164.
Leicester co-operative builders, 111, 287.
   engineers, 118, 142.
   "Equity," 100, 106, 132, 276, 285, 330.
   hosiery, 113.
   printers, 119.
Libraries, co-operative, 148, 155.
Lincoln farm, 20, 21, 25.
   society, 78.
Livesey, George, 191.
   in *International Review*, 207.
Livesey, Thomas, 191.
Loans, government, to individuals, 68.
London builders, 147, 237.
   co-operation in, 235, 336.
Loss sharing, 154, 224.
Losses of farming, 18, 41.
Ludlow, J. M., 4, 216, 231, 246, 262, 268, 278, 298, 336.
Luzzatti, 336.

Magee, Archbishop, 296.
Managers, co-operative, compensation of, 261, 272.
Mann, Tom, 194, 231.
Manure Manufacturers' Alliance, 63.
Market, co-operative, 149, 269, 274, 308.
Marshall, Professor, 231, 284, 295.

Maurice, F. D., 4, 216, 235, 297, 335, 336.
Maxwell, William, 291, 320.
McInnes, D., 16, 21, 83.
McLeod, Alexander, 28.
Methodists, 16.
Migration to towns, 80.
Mill, J. S., 295.
Minneapolis flour mill consolidation, 267.
Mitchell, Hey, mill, 276.
Molijn, M., 7.
Montgomery, J. K., 67.
Moorhouse, Bishop, 285.
Morley, Lord, 204.
Music, 243.

Neale, E. V., 4, 170, 171, 183, 216, 217, 223, 309, 336.
Nelson, N. O., 199.
Newcastle farm, 22.
   picnic, 80.
New Lanark, 294.
News, The Co-operative, 45, 209, 332.
Notes, labor, 88.
Nottingham, 10, 13.
Nunspeet, 7.
Nuttall, 105.

Owen, Robert, 4, 98, 215, 222, 294, 297, 336.

Paisley, 166.
Pall Mall Gazette, 210.
Panic of 1894, 30.
Paris International Congress, 280.
   international store, 5.
   municipal co-partnership, 317.
Parnell, 53.
Parson and co-operation, 48, 49.
People's banks, 8, 52, 64, 242.
Petersborough, 13.
Piece-work, 136.
Playground, co-operative, 247.
Plunkett, Horace, 53, 54, 58, 59, 62, 66, 69, 72, 76, 225, 253, 336.

Politics of co-operators, 13, 91, 315, 316.
Pooling of profits, 39, 92, 299.
Population, co-operative, 1, 130.
Potter, John, 106.
Printing, co-operative, 7, 119, 263.
Productive federation, 240.
Profits, 31.
 division of, 110, 119, 173, 222, 224.
 Irish creameries, 63.
 pooled, 39, 92, 299.
Profit-sharing, English Co-operative Wholesale, 276.
 at Rochdale, 276.
 South Metropolitan Gas Co., 194.
 Wm. Thomson & Sons, 293.
Public contracts and labor copartnership, 317.
Publicity, 268.
Pullman, 299.
Puritans, 15.

Raffeisen, 336.
Railroads, 15, 77, 95, 253.
Ralahine, 86.
Reading rooms, 155, 308.
Recess committee, report of, 55, 56.
Recreation rooms, co-operative, 107, 307.
Redemption societies, 278.
Rents, 13, 43, 45, 95.
Representation of employés in management, 47, 292, 313 (see employés).
Right to work, 113.
Ring in manures, 63.
Ripon, Marquis of, 4, 49, 170.
Ritchie, 85.
Rochdale, 4, 90, 97, 178, 220, 276, 282, 294.
Rogers, Thorold, 295.
Roseberry, Lord, on co-operation, 17.
Royal Arsenal Co-operative Society, 28.
Rugby, 9.
Russia, 62.

Salford store, 278.
 boycott, 252.
Schultze-Delitzsch, 295, 336.
Scotland co-operation, 275.
Scottish Co-operative Farming Association, 20, 41.
Scottish Co-operative Wholesale Society, 19, 20, 26, 45, 71, 181, 273, 275, 288, 299, 330, 341.
Seaside Home, 318.
Self-governing workshop, 215, 222, 234.
Self-help, 3, 142.
 by each other help, 145, 282.
 Ireland, 59.
Selling and buying, Irish societies, 61.
Sentiment, 11, 65, 123, 131, 137, 139, 150, 188, 189, 250, 270, 327.
Separatists, 15.
Shareholders (see employés).
Shieldhall, 300.
Shirtmaking, 302.
Sick benefits, 192.
Sliding scale, bonus to labor, 196.
 dividends, 193.
Soap works, 280, 281, 306.
Socialism, Canadian creamery, 75.
 and co-operation, 185, 300.
Socialist, The Christian, 216, 278.
Song of the Shirt, 304.
South Metropolitan Gas Co., 191.
Speculation in co-operative shares, 144, 180.
Stamford, Earl of, 231.
Store, International Co-operative, 5.
Stores, co-operative, against co-operative production, 122, 155.
Strike, gas works, 197.
Strikes, 100, 104, 106, 107, 136, 167, 170, 266.
Subsidized co-operation, 224, 242.
Superannuation, 192.
Suspension of employés, 67, 104.
Sweater, competition of co-operation with, 302.
Sweating, charges of, 284, 343.

Tennyson, 96.
Thomson, Daniel, 23.
Thomson, George, 266, 269, 293.
Thomson, William & Sons, 293.
Thurles Conference, 69.
Tillett, Ben., 194.
Tisserand, M., 73.
Tithes, 15.
Trades-unionists and South Metropolitan Gas Co. bonus, 197.
Trades-unions, 104, 107, 108, 109, 116, 135, 136, 137, 168, 187, 199, 210, 236, 238, 301, 308, 312, 336, 342.

United Co-operative Baking Society, 47, 311.
United States Agricultural Department, 73.
Universal suffrage in business, 134.

Values, forces determining, 123.
Van Marken, J. C., 6, 7, 336.
Vernon, Lord, 49.
Village communities, 7, 94, 97, 215, 334.
Visits, co-operative, 248.
Vivian, Henry, 9, 82, 140, 150, 215, 234, 237, 331.
Wage-balance fund, 158.

Wages, 66, 107, 108, 136, 168, 186, 229, 236, 238 (see agricultural laborer) (see dividend on wages).
Walsall Padlock Makers, 343.
Webb, Mrs. Sidney, 55, 260, 264, 273.
Wesleyans, 15, 16.
Westcott, Bishop, 296.
West Killbride Seaside Home, 318.
Wheat Sheaf Shoe Works, 100, 103, 285.
Williams, Aneurin, 215, 231.
Williams, John, 164.
Winchilsea, Earl of, 80, 84, 92, 259.
Withdrawable capital, 30, 269.
Wolff, Henry W., 8, 242.
Women, 3, 109, 134, 136, 144, 149, 151, 152, 185, 186, 202, 300.
farm wages of, 38.
Women's Co-operative Guild, 248, 320, 324.
Woolwich farm, 19, 20, 28.
Workingmen as business men, 84, 109, 152, 251, 302.
Workmen directors, 184, 191 (see employés).
Workmen's Times, 114.
Workshop, self-governing, 215, 222, 234.

By HENRY MILLS ALDEN

A STUDY OF DEATH. Post 8vo, Half Leather, Uncut Edges and Gilt Top, $1 50.

The boldly imaginative beauty, the insight into spiritual realities, and the mystic temper . . . make it one of the most remarkable works in the field of ethics and psychology produced in America.—*Hartford Courant.*

The work of a thinker. Its intellectual power holds, its argument compels. It is destined to be a book of indefinitely extended service for this reason. It is a book wherein a thousand ministers may find interpretation of the Biblical words of comfort, of the promise of joy. But it is vastly more than a book for preachers.—*Boston Transcript.*

GOD IN HIS WORLD. An Interpretation. Book I. From the Beginning. Book II. The Incarnation. Book III. The Divine Human Fellowship. Post 8vo, Cloth, Uncut Edges and Gilt Top, $1 25; White and Gold Edition, $2 00.

A prose poem, in fact, inspired by reverence for God and religion, and which traces from the dawn of history "the prophecy, antitype, and fulfilment of the coming of Christ." Following what he considers a regular law of human development, he constructs a theistic system which will fascinate some readers and interest many more who may not necessarily accept his opinions. It is a purely individual treatise, in no respect controversial, in which Christ takes the central place both in the Gospel revelation and in all true explication of nature and society.—*N. Y. Sun.*

NEW YORK AND LONDON
HARPER & BROTHERS, PUBLISHERS

By HENRY LAUREN CLINTON

CELEBRATED TRIALS. With Nine Portraits. Crown 8vo, Cloth, Uncut Edges and Gilt Top, $2 50.

The author has left no need to write up his subject. He marshals the preliminary facts in each case clearly and dispassionately, and then lets the story in a great measure tell itself. The author has relied to a considerable extent upon extracts from the newspapers of the day, to which his own matter supplies the links as well as a running commentary. The effect of this method is that his pictures have the old-time coloring and atmosphere, and we see the events, as it were, in their proper perspective.—*San Francisco Bulletin.*

The stories will be read for their own absorbing interest, as well as for the light they throw on municipal history.... We are given facts untouched by fancy, and the stories are interesting enough in themselves to hold the attention from beginning to end.—*Saturday Evening Gazette*, Boston.

EXTRAORDINARY CASES. With Photogravure Portrait. Crown 8vo, Cloth, Uncut Edges and Gilt Top, $2 50.

The number and importance of the cases in which Mr. Clinton was interested is indeed extraordinary, and their description has for even the unprofessional reader a fascinating interest.... Mr. Clinton's book is interspersed with interesting anecdotes of bench and bar, and cannot fail to interest lawyer and layman alike.—*N. Y. Mail and Express.*

NEW YORK AND LONDON
HARPER & BROTHERS, PUBLISHERS

By GEORGE DU MAURIER

THE MARTIAN. A Novel. Illustrated by the Author. Post 8vo, Cloth, Ornamental, $1 75; Three-quarter Calf, $3 50; Three-quarter Crushed Levant, $4 50. (A Glossary of the French and Latin expressions in the story is included.)

> The romance has the ring of Mr. Du Maurier's best romancing; the simple, almost naïve, admiration of the boys for Barty shows the author as we have known him in his highest estate—true, wise, free from the slightest suspicion of sentimentality or cant. "The Martian" opens again the portals of his delightful world, the story revives the tenderness, the sweetness, the original magic which many readers have feared could never be recaptured—*N. Y. Tribune.*

SOCIAL PICTORIAL SATIRE. Reminiscences and Appreciations of English Illustrators of the Past Generation. With Illustrations by the Author and Others. Post 8vo, Cloth, Ornamental, $1 50.

A LEGEND OF CAMELOT. Pictures and Verses. Oblong 4to, Cloth, Ornamental, Full Gilt, $5 00. (*In a Box.*)

TRILBY. A Novel. Illustrated by the Author. Post 8vo, Cloth, Ornamental, $1 75; Three-quarter Calf, $3 50; Three-quarter Crushed Levant, $4 50.

PETER IBBETSON. With an Introduction by his Cousin, Lady **** ("Madge Plunket"). Edited and Illustrated by GEORGE DU MAURIER. Post 8vo, Cloth, Ornamental, $1 50. Three-quarter Calf, $3 25; Three-quarter Crushed Levant, $4 25.

ENGLISH SOCIETY. Sketched by GEORGE DU MAURIER. About 100 Illustrations. With an Introduction by W. D. HOWELLS. Oblong 4to, Cloth, Ornamental, $2 50.

NEW YORK AND LONDON
HARPER & BROTHERS, PUBLISHERS

By FRIDTJOF NANSEN

FARTHEST NORTH. Being the Record of a Voyage of Exploration of the Ship *Fram* (1893-1896), and of a Fifteen Months' Sleigh Expedition by Dr. NANSEN and Lieut. JOHANSEN. With an Appendix by OTTO SVERDRUP, Captain of the *Fram*. With over 100 Full-page and Numerous Text Illustrations, Sixteen Colored Plates in Facsimile from Dr. NANSEN's own Water-Color, Pastel, and Pencil Sketches, an Etched Portrait, Two Photogravures, and Four Maps. pp. xxvi., 1301. Two Volumes, Large 8vo, Uncut Edges and Gilt Tops, $10 00 ; Half Leather, $12 50.

Stanley's "Through the Dark Continent" is the only work of recent years that can compare with Nansen's in importance, daring, and adventure.—*Chicago Tribune.*

Merely to turn the leaves of the two handsome volumes is to see what a wealth of scientific and personal interest they contain. The illustrations are of the highest value—most of them being after Nansen's own photographs.—*N. Y. Evening Post.*

It is a story that will live through age after age.—*London Chronicle.*

It is not too much to say that the book is a masterpiece of story-telling.—*London Times.*

Not more than once in a generation, if as often as that, is such a narrative presented to the world.—*N. Y. Tribune.*

Thrilling adventures, hairbreadth escapes, and magnificent sport with Arctic monsters.—*N. Y. Herald.*

FARTHEST NORTH. Popular Edition. One Volume. With Sixteen Illustrations. 8vo, Cloth, $2 50.

NEW YORK AND LONDON
HARPER & BROTHERS, PUBLISHERS

www.ingramcontent.com/pod-product-compliance
Lightning Source LLC
Chambersburg PA
CBHW022333230426
43664CB00040B/448